Isolation and characterization of new psychrophilic bacteria; Cloning and characterization of industrial relevant pullulanase and serine protease

Dem Promotionsausschuss der

Technischen Universität Hamburg-Harburg

zur Erlangung des akademischen Grades

Dr. rer. nat.

genehmigte Dissertation

Von

M.Sc.-Biol. Farah Qoura

aus Jordanien

Hamburg 2006

Die vorliegende Arbeit wurde im Arbeitsbereich Technische Mikrobiologie der Technischen Universität Hamburg-Harburg durchgeführt

Vorsitzender des Prüfungsausschusses: Prof. Dr.-Ing F. Keil
1. Gutachter: Prof. Dr. rer. nat. G. Antranikian
2. Gutachter: Prof. Dr. rer. nat. R. Müller

Tag der mündlichen Prüfung: 07.04.2006

Bibliografische Information Der Deutschen Bibliothek

Die Deutsche Bibliothek verzeichnet diese Publikation in der Deutschen Nationalbibliographie; detallierte bibiografische Daten sind im Internet über http://dnb.ddb.de abrufbar.

Dissertation (Ph.D. thesis), Technische Universität Hamburg-Harburg

Printed in Germany
ISBN 3-8334-4967-5

Herstellung und Verlag: Books on Demand GmbH, Norderstedt
Gedruckt auf alterungsbeständigem Papier nach ISO 9706 (säure-, holz- und chlorfrei).

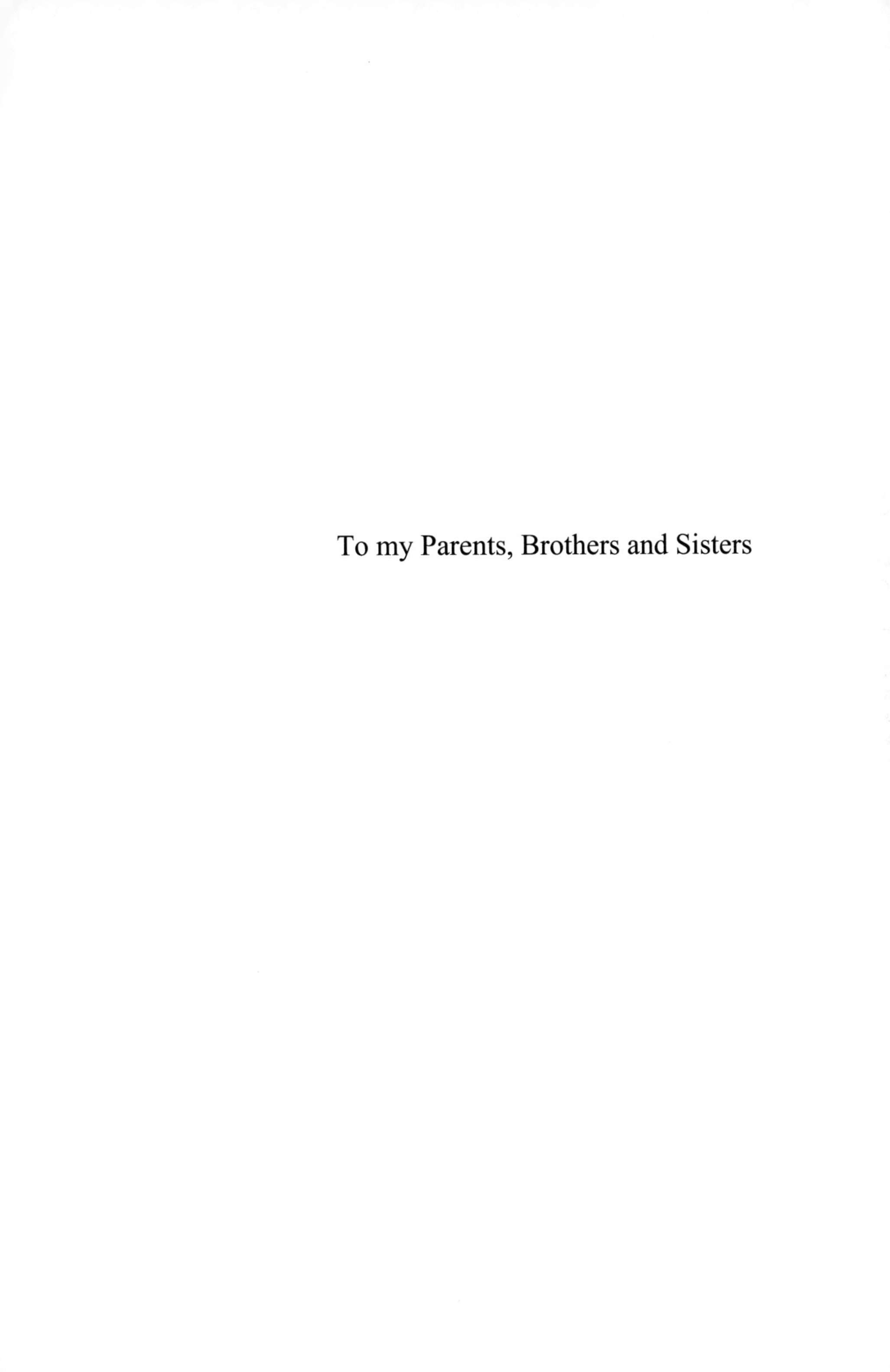

To my Parents, Brothers and Sisters

TABLE OF CONTENT

1 INTRODUCTION

1.1 The revolution of biotechnology

"Why trouble to make compound yourself when a bug will do it for you?"
J.B.S. Haldane, 1929 (Prentis, 1989)

More than five decades ago (1929) Haldane, one of the most perceptive scientists of his time, established the logic behind what is now called biotechnology (Prentis, 1989). Biotechnology is one of the most used words in modern biology (Brown *et al.*, 1988). It has been defined in many ways, such as the application of biological organisms, systems or processes to manufacture and to service industries (Smith, 1990) - the controlled and deliberate application of simple biological agents - living or dead cells or cell components - in technically useful operations, either of productive manufacture or as service operations (Bu'lock & Kristiansen, 1987). In other words, biotechnology is of an interdisciplinary nature and requires the integration of biochemistry, biology, microbiology, chemical engineering and process engineering together with other disciplines in a way that optimize the exploitation of their potential (Bu'lock & Kristiansen, 1987; Smith, 1990). European Federation of Biotechnology (EFB) defined biotechnology as the integration of natural sciences and engineering in order to achieve the application of organisms, cells, parts thereof and molecular analogues for products and services (EFB, 1989).

Biotechnology has been known and applied for thousands of years. Before and during 6000 BC, wine, and beer were produced by the aid of microbes (yeast) (Prentis, 1989; Smith, 1990). In 4000 BC leavened bread was produced also by the aid of yeast (Prentis, 1989; Smith, 1990). In the year 1521 algae were harvested from lakes by the Aztecs and they were used for food (Smith, 1990).

In the late 16th and early 17th century, Antoni van Leeuwenhoek observed microbes with his newly designed microscope for the first time, which became the first step toward the knowledge of biotechnology. Late during the 18th century fermentative ability of microorganisms was demonstrated by Louis Pasteur the father of biotechnology. Eduard Buchner, showed that the enzymes extracted from yeast could convert sugar into alcohol (Prentis, 1989; Smith, 1990).

In the first half of the 19th century large-scale sewage purification systems employing microbes were established, Alexander Fleming discovered penicillin, and double helix structure of DNA was revealed. Later in the 19th century and during the seventies, genetic engineering experiments were followed, as the beginning of applied genetics and recombinant DNA technology. In addition, the technology for the production of monoclonal antibodies was then created (Prentis, 1989; Smith, 1990).

During the last decades, biotechnology has been developing very fast and could be applied in all aspects of life. The importance of biotechnology lies in: the production of food crops, livestock husbandry and animal health, pharmaceutical and chemical industries, the accurate diagnosis and prevention of human diseases, the conversion of biomass into energy, the transformation of wastes and agricultural and industrial bio-products, pollution control and environmental sanitation (Bull *et al.*, 1982; Hacking, 1987; Yanchinski, 1985; Zimmerman, 1984a,b).

1.2 Extremophiles

"Once upon a time, all of life was extremophiles-living on the edge"
(David, 1999)

Once upon a time, all of life was extremophiles, that is, living under extreme conditions—living "on the edge". Extreme conditions can refer to not only physical extremes (e.g. temperature, pressure or radiation), but also to geochemical extremes such as salinity and pH. Most of the extremophiles that have been identified to date belong to the domain of the Archaea. However, other extremophiles from the bacterial and eukaryotic kingdoms have also been recently identified and characterized (Fig.1-1) (Gomes & Steiner, 2004; van den Burg, 2003).

Extremophiles are unique microorganisms that are adapted to survive in ecological niches such as high or low temperatures, extremes of pH, high salt concentration and high pressure. Accordingly, biological systems and enzymes can even function at temperatures between -5 and 130 ° C, pH 0-12, salt 3-35 % and 1000 bar (Antranikian *et al.*, 2005). Extremophiles are classified on the base of extreme environmental conditions as follows:

1. *Low-temperature-adapted microorganisms*: They can be divided into two main groups: psychrophiles and psychrotolerants. Psychrophiles grow at an optimum temperature of 15 °C, with a maximum growth temperature at about 20 °C and a minimum around 0 °C. Psychrotolerants generally do not grow at 0 °C but do so at 3-5 °C, and have optimum and maximum growth temperature above 20 °C but less than 30 °C. *Vibrio sp, Micrococcus criophilus* and *Aquaspirillum articum* are some examples of low-temperature-adapted microorganisms (Antranikian *et al.*, 2005; Gomes & Steiner, 2004).

2. *Thermophiles*: They can be divided into three groups: moderate thermophiles which grow optimally between 50 and 60 °C, such as *Bacillus acidocaldarius*, extreme thermophiles which grow optimally between 60 and 80 °C, such as *Thermus aquaticus,* and finally hyperthermophiles which grow optimally between 80 and 108 °C, such as *Thermotoga maritima* (Antranikian *et al.*, 2005).

3. *Alkalithermophiles/Alkaliphiles*:Alkalithermophilic microorganisms grow optimally under two extreme conditions: at pH values between 8 and 10 or above, and at high temperatures (50-85 °C), such as *Anaerobranca gottschalkii* (pH 9 and 85 °C). On the other hand, microorganisms classified as alkaliphiles are mesophiles and consist of two main physiological groups: alkaliphiles and haloalkaliphiles. Alkaliphiles grow aptimally at around pH 10, whereas haloalkaliphiles require both an alkaline pH (pH> 8) and high salinity (up to 33 % NaCl) (Gomes & Steiner, 2004).

4. *Acidothermophiles/Acidophiles*: Acidothermophiles thrive under conditions of low pH and high temperatures. For example, the acidothermophile *Sulfolobus solfataricus* grows at pH 3 and 80 °C. True acidophile such as archaea *Picrophilus torridus,* grow optimally at pH values as low as 0.7 and at 60 °C (Gomes & Steiner, 2004).

5. *Halophiles*: Halophiles comprise bacteria and archaea that grow optimally at NaCl concentration above that of seawater (> 0.6 M NaCl) such as *Clostridium halophilum.* In general, halophiles are classified as moderate halophiles if they can grow at salt concentration between 0.85 and 1.7 M NaCl and as extreme halophiles if they require NaCl concentration above 1.7 M for growth (Antranikian *et al.*, 2005).

6. *Piezophiles*: Microorganisms that like high pressure conditions (38-130 MPa) for growth are termed piezophiles (formerly known as barophiles). Piezophiles are distributed among different genera, such as *Shewanella* and *Moritella* (Gomes & Steiner, 2004).

7. *Radiophiles*: Microorganisms that are highly resistant to high levels of ionizing and ultraviolet radiation (3-5 Mrad) are called radiophiles. Examples of radiophiles are *Deinococcus radiodurans* and *Thermococcus radiotolerans* (Gomes & Steiner, 2004).

8. *Metallophiles*: Microorganisms that can grow in the presence of high metal iones concentrations are called metallophiles.

Ralstonia metallidurans is an example of a metallophile (Gomes & Steiner, 2004).

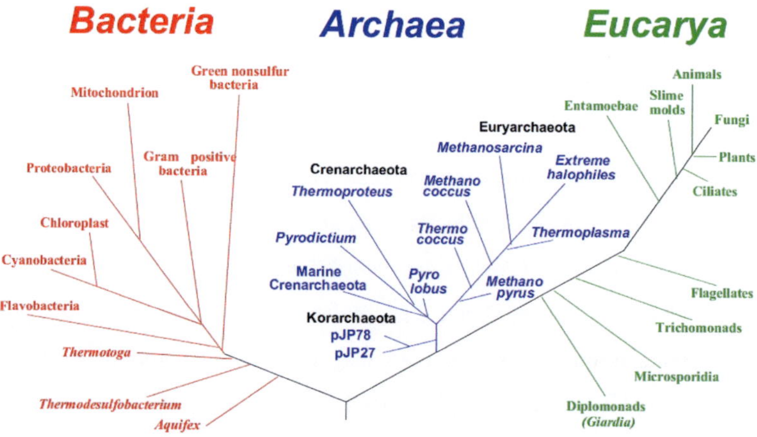

Fig. 1-1: Rooted universal phylogenetic tree showing the three domains based upon 16S (or 18S) rRNA sequences. The position of the root was determined by comparing paralogous gene sequences that diverged from each other before the three primary lineages emerged from their common ancestral condition. (Madigan *et al.*, 2000)

1.2.1 Habitat of extremophiles

A large part of Earth's surface is characterized by low temperatures. In the oceans that cover 71% of the planet surface the average yearly temperature is 5 °C and in the ocean depths the temperature is between 1 and 4°C throughout the year due to the combined effects of hydrostatic pressure and water density. Polar regions, representing 14 % of the surface of the Earth, are permanently frozen, or the temperature above freezing point is only for a short period of time. These environments, which are dominant on Earth, are favorable to psychrophiles and psychrotolerants able to grow at any cold temperature at which water is still liquid (Madigan *et al.*, 1997)

Seawater contains on average 3% NaCl (w/v), and salt concentration in water above 3% are rare in nature. The Dead Sea and the Great Salt Lake in the USA are large natural hypersaline environments on Earth, whereas the Blue Lagoon in Iceland and the solar salterns in Western France are few examples of typical man-made saline or hypersaline ecosystems. Moderate halophiles grow in NaCl concentration above of seawater (> 3 %) and extreme halophiles can grow in saturated NaCl (>30%) (Grant, 1991). Furthermore, saline lakes contain high amounts of dissolved organic matter and are regarded as such as extremely productive environments (Grant, 1991).

Extreme pH environments restrict the occurrence of microbial life. Acid environments are widely distributed in food and soils all over the world. Large geographic zones rich in sulfur compounds are linked to volcanic activity. Hydrogen sulfide (H_2S) is released in large amounts and oxidized chemically and biologically into sulfuric acid, resulting in the decrease of the pH in the soil. On the other hand, there are few examples of highly alkaline biotopes in nature. Soda lakes in West-Africa, Tibet, China or California have pH values reaching 11 to 12 and also contain high concentrations of salts. Unlike natural hypersaline lakes and seas, they are depleted in Ca^{2+} and Mg^{2+} ions which disappeared at the early stages of the creation of the lake by precipitation of carbonates (Grant, 1991).

Extreme emissions of radiation are mostly found in man-made nuclear facilities. Very few microorganisms have been identified, among these *Deinococcus radiodurans* that can survive as much as 3 to 5 Mrad when a lethal dose for humans is 0.0001 Mrad, Recent research indicates that its ability to resist radioactivity may results from the presence of multiple copies of the chromosome and the ability to repair severely damaged DNA (Prescott *et al.*, 1999).

High-temperature environments are generally found associated with volcanic activity and in man-made industrial complexes. The most important biotopes are terrestrial geothermal fields with alkaline freshwater hot springs and solfatara, and marine environments with coastal, shallow and deep hydrothermal systems. Hot environments display a complete range of pH, from acid to alkaline, depending on temperature, water availability, and gases and ion concentration (Kistjansson & Herggvidsson, 1995).

1.2.2 Extremophiles as a source of novel enzymes

Due to the increasing industrial demand for biocatalysts that can cope with industrial process conditions, considerable efforts have been devoted to the search for such enzymes. Compared with organic synthesis, natural biocatalysts often have much better chemical precision, which can lead to more efficient production of single stereoisomers, fewer side reactions and a lower environmental burden (Rozzell, 1999). The notion that extremophiles are capable of surviving under non-standard conditions in non-conventional environments has led to the assumption that the properties of their enzymes have been optimized for these conditions. Indeed, data for a considerable fraction of the enzymes that have been isolated and functionally characterized from extremophiles support this assumption (Gomes & Steiner, 2004; Hough & Danson, 1999; van den Burg, 2003).

The classification of 'extreme environments' refers to a wide variety of different conditions to which microorganisms have adapted. The biocatalysts obtained from these microorganisms could be applied in similarly diverse conditions (Table 1-1). For the degradation of polymers such as chitin, cellulose or starch, enzymes that are active at and resistant to, high temperatures are often preferred. Under these conditions the solubility and, consequently, the accessibility of the substrate is improved. Alternatively, if one needs to perform a stereospecific modification of a compound for the synthesis of a pharmaceutically relevant product in organic solvents, very different prerequisites apply to the biocatalysts. As salt is known to reduce water activity, enzymes from halophilic microorganisms could be the most suitable choice for application in nonaqueous media (Marhuenda-Egea *et al.*, 2002). This diversity of environments to which different extremophiles have adapted offers many exciting opportunities for a variety of applications (van den Burg, 2003).

Table 1-1: Classification of extremophiles and examples of applications of some of their enzymes.

Type	Growth characteristics	Enzymes	Applications
Thermophiles	Temp >80 °C	Proteases	Detergents, hydrolysis in food and feed, brewing, baking
		Glycosyl hydrolases (e.g. amylases,	Starch, cellulose, chitin, pectin
		pullulanase, glucoamylases, glucosidases, cellulases, xylanases)	processing, textiles
		Chitinases	Chitin modification for food and health products
		Xylanases	Paper bleaching
		Lipases, esterases	Detergents, stereo-specific reactions (e.g. trans-esterification, organic biosynthesis)
		DNA polymerases	Molecular biology (e.g. PCR)
		Dehydrogenases	Oxidation reactions
Psychrophiles	Temp < 15 °C	Proteases	Detergents, food applications (e.g. dairy products)
		Amylases	Detergents and bakery
		Cellulases	Detergents, feed and textiles
		Dehydrogenases	Biosensors
		Lipases	Detergents, food and cosmetics
Halophiles	High salt, (e.g. 2–5 M NaCl)	Proteases	Peptide synthesis
		Dehydrogenases	Biocatalysis in organic media
Alkaliphiles	pH >9	Proteases, cellulases	Detergents, food and feed
Acidophiles	pH >0.7	Amylases, glucoamylases	Starch processing
		Proteases, cellulases	Feed component
		Oxidases	Desulfurization of coal
Piezophiles	High pressure; up to130 MPa	To be defined	Food processing and antibiotic production

Table after van den Burg, 2003

1.3 Psychrophilic bacteria

"Greek words: *psychros* and *philos* making cold-loving or psychrophiles"
(Ingraham & Stokes, 1959)

Psychrophiles is a term derived from the Greek word psychros, meaning cold, and philos, meaning loving, i. e., cold-loving. They have been called also cryophiles and rhigophiles which both are also derived from the Greek and have essentially the same meaning as psychrophile (Ingraham & Stokes, 1959). To the best of our knowledge, the term psychrophile was used first used by Schmidt-Nielsen in 1902 for microorganisms that can grow at 0 °C, although such bacteria were described first by Foster in 1887 (Ingraham & Stokes, 1959). Schmidt-Nielsen (1902) defined psychrophiles as organisms that are not only able to survive, but also multiply at 0 °C (Ingraham & Stokes, 1959). Morita (1975) defined psychrophiles as organisms having an optimal temperature for growth at about 15 °C or lower. Recently, psychrophilic microorganisms are defined as the microorganisms that grow at an optimum temperature of 15 °C with a maximum growth temperature at about 20 °C and a minimum around 0 °C. Microorganisms that do not grow at zero but do so at 3-5 °C, and have optimum and maximum growth temperature above 20 °C but less than 30 °C are defined as psychrotolerant microorganisms (Antranikian *et al.*, 2005). Many psychrophiles live in biotopes having more than one stress factor, such as low temperature and high pressure in deep seas (piezo-psyhrophiles), or high salt concentration and low temperature in the sea ice (halo-psychrophiles) (Gomes & Steiner, 2004).

1.3.1 Isolation and habitat of psychrophilic bacteria

Cold adapted microorganisms are found in both permanently and temporarily cold habitats, which comprise more than 80% of the Earth's biosphere. Oceans, high mountains and deep lakes provide various aquatic and terrestrial cold environments where the temperature seldom or never reaches 5 °C (Gounot, 1999). Many psychrophilic and psychrotolerant bacteria, which grow between 0 and 30 °C, have been previously isolated from the surface and deep-sea environments (Maruyama *et al.*, 2000). However, these low-temperature bacteria, are widely distributed and occur in large numbers in soil, water, foods, and other habitats where they carry out important beneficial or harmful chemical transformations at temperatures which are too low for the growth of other types of bacteria (Ingraham & Stokes, 1959).

Early historical investigation on psychrophily reported that Forster in 1887 (Berry & Magoon, 1934) was the first to demonstrate the ability of pure cultures of bacteria to grow at 0 °C. Forster's cultures were luminous bacteria isolated from marine fish. Within a year, Forster's results were confirmed by Fischer (Berry & Magoon, 1934) with 2 strains of luminous bacteria in his culture collection and also with 14 additional strains of freshly isolated marine bacteria. In a subsequent paper in 1892 (Ingraham & Stokes, 1959; Morita, 1975), Forster showed that psychrophilic bacteria are widely distributed in nature. He found them in fresh and salt water, on the surface and in the intestines of fresh and salt water fish, in milk and meat, in garden soil and street dirt, and in canal and meadow ditch water. Early investigators reporting on the identification of psychrophilic bacteria are listed in Table 1-2.

Table.1-2: Early investigators reporting on the occurrence of psychrophilic bacteria

Investigator and year	Source material or organism
Forster (1887, 1892)	Fish, natural waters, foods, wastes, rubbish, soil, surface and intestines of fish
Fischer (1888, 1893)	Harborwater and soils, certain pathogenic bacteria
Conradi and Vogt (1901)	"proteus bacillus"
Baur (1902)	*Bacterium lobatum*
Brandt (1902)	*Bacterium lobatum*
Schmidt-Nielsen (1902)	*Bacillus aquatile fluorescense non-liquefaciens, B. granulosum, B. paracoligasoformans anindolicum, B. radiatum, B. tarde fluorescens, B. pestis,and B. proteus fluorescens.*
Miller (1903)	Sausage meat, fish, intestinal contents of fish, milk, vegetables, meal, garden soil and muck
Feitel (1903)	Dentrifying bacteria from deep sea water
Tsiklinsky (1908)	Antartic fish
Richardson (1908)	Nonproteolytic bacteria on frozen meat
Pennington (1908)	Milk
Ravenel et al. (1910)	Milk
Conn (1910,1912, and 1914)	Frozen soils
Brown and Smith (1912)	Frozen soils
Reed and Reynolds (1916)	Milk
Vanderleck (1917 and 1918)	Frozen soils
Vass (1919)	Frozen soils
Fay and Olson (1924)	Ice cream
Bardach (1924)	Sewage, sludge and salt water
Rubentschik (1925)	*Urbacillus psychrocartericus, Urosarcina psychrocartericus*

Table after Morita, 1975

1.3.2 Psychrophiles as a source of novel cold-adapted enzymes

Psychrophilic organisms live at low temperatures, where most other species cannot grow and survive. To do so, they need to produce enzymes that are able to perform their catalysis efficiently under these extreme environmental conditions (Gianese *et al.*, 2001). At the same temperature, enzymes from mesophilic or thermophilic organisms are generally unable to sustain a viable metabolism (Gerday *et al.*, 2000; Gianese *et al.*, 2001). For these reasons, enzymes synthesized by psychrophilic organisms have considerable biotechnological potential. Their ability to work efficiently as biocatalysts at low temperatures offers advantages in environmental applications and in saving energy when they are used in industrial processes (Gianese *et al.*, 2001). Some psychrophilic enzymes are shown in Table 1-3.

Table.1-3: List of some cold-active enzymes from different psychrophiles

Family name	Species	Growth temp. [°C]
Alanine racemase	*Bacillus psychrosaccharolyticus*	15
α-Amylase	*Alteromonas haloplanctis*	4
Aspatate carbamoyltansferase (catalytic chain)	*Vibrio* sp.	6
Aspatate carbamoyltansferase (regulatory chain)	*Vibrio* sp.	6
β-Galactosidase	*Arthrobacter* sp.	15
β-Lactamase	*Psychrobacter immobilis*	4
Chymotrypsin A	*Gadus morhua*	4
Citrate synthase	*Antarctic bacterium*	5
DNA ligase	*Pseudoalteromonas haloplanktis*	4
Elastase	*Salmo salar*	4
Isocitrate dehydrogenase 1	*Vibrio* sp.	4
3-Isopropylmalate dehydrogenase	*Vibrio* sp.	15
L-Lactate dehydrogenase P	*Bacillus psychrosaccharolyticus*	15
Malate ehydrogenase	*Aquaspirillum arcticum*	4
Ornithine carbamoyltransferase	*Vibrio* sp.	6
Pyruvate kinase	*Bacillus psychrophilus*	15
Serralysin (Alkaline protease)	*Pseudomonas aeruginosa*	4
Subtilisin	*Bacillus* sp.	4
Triosephosphate isomerase	*Vibrio marinus*	15
Trypsin I	*Salmo salar*	4
Xylanase	*Cryptococcus adeliae*	4

Table after Gianese, *et al.*, 2001

The application of these enzymes offers considerable potential to the biotech industry, for example, in the detergent and food industry, for the production of fine chemicals and in bioremediation processes. If today the annual market for thermostable enzymes represents ~US$250 million, it is likely that the potential value of cold-adapted enzymes is greater in view of the diverse capabilities of these enzymes. These offer significant advantages at the level of their specific activity, lower stability and unusual specificity (Gerday *et al.*, 2000; Gianese *et al.*, 2001). Some of the more interesting applications (Gerday *et al.*, 2000) include the following:

1. *Detergent additives*: A well known and important application of enzymes, such as proteases, lipases, α-amylases and cellulases, is their use as additives in detergents. A reduction in energy consumption and a reduction in wear and tear are obvious advantages.

2. *Textile industry*: Cold-adapted cellulases are used for biopolishing and stone-washing processes.

3. *Food industry*. In the milk industry, β-galactosidase is used at low temperature to reduce the amount of lactose responsible for severe induced intolerances in approximately two thirds of the world's population. The use of pectinases helps the juice extraction process, reduces the viscosity and helps to clarify the final product. In the meat industry, proteases help to tenderize the meat. In baking processes, enzymes such as amylases, proteases and xylanases can be used to reduce the dough fermentation time, improve the properties of the dough and the crumb, in addition to the retention of aromas and moisture levels.

1.4 Pullulanases

"Pullulanase is a microbial enzyme, breaks down branched polysaccharides"
(MacGregor, 2003)

Pullulanase (EC 3.2.1.41) catalyzes hydrolysis of α- 1,6-D-glucosidic linkages in pullulan, amylopectin, and the amylose. The name pullulanase has been used for the microbial enzymes important for the breakdown of branched polysaccharides where, the names limit dextrianse and R-enzyme have been applied to enzymes of plant origin (MacGregor, 2003).

According to Kuriki & Imanaka (1999)(Kuriki & Imanaka, 1999), GH13 enzymes are defined by (1) the hydrolysis of α-glycosyl bonds, (2) the hydrolysis or transfer of α-glycosyl bonds by retention of the anomeric conformation, (3) the conservation of four amino acid sequence motifs containing the catalytic residues and most other residues involved in substrate-binding within the catalytic site, and (4) the presence of Asp, Glu and Asp as catalytic residues. The highly conserved residues are located within the motifs of the catalytic domain and additional non-catalytic modules from family GH13 (Fig. 1-2). They are used for the classification of the enzyme within the family and can consequently serve for the design of degenerate oligonucleotide primer for PCR (Kuriki & Imanaka, 1999).

	I (β2)	II (β4)	III (β5)	IV (β7)
Amylomaltase	EALGIRIIGDMPIFVAED	LFHLVRIDHFRG	VPVLAEDLGVI	VVYTGTHDNDT
Amylosucrase	HEAGISAVVDFIFNHTSN	GVDILRMDAVAF	VFFKSEAIVHP	VNYVRSHDDIG
CGTase	HAKNIKVIIDFAPNHTSP	GIDGIRMDAVKH	VFTFGEWFLGV	VTFIDNHDMER
Cyclomaltodextrinase	HDNGIKVIFDAVFNHCGY	DIDGWRLDVANE	AIIVGEVWHDA	FNLIGSHDTER
Branching enzyme	HQAGIGVILDWVPGHFCK	HVDGFRVDAVAN	ILMIAEDSTDW	FILPFSHDEVV
Isoamylase	HNAGIKVYMDVVYNHTAE	GVDGFRFDLASV	LDLFAEPWAIG	INFIDVHDGMT
Maltogenic α-amylase	HQKAIRVMLDAVFNHSGY	DIDGWRLDVANE	AYILGEIWHDA	FNLLGSHDTPR
Pullulanase	HAHGVRVILDGVFNHTGR	GVDGWRLDVPNE	AYIVGEIWEEA	MNLLTSHDTPR
Sucrose phosphorylase	LGECSHLMFDFVCNHMSA	GAEYVRLDAVGF	TVIITETNVPH	FNFLASHDGIG
α-Amylase	HERGMYLMVDVVANHMGY	SIDGLRIDTVKH	VYCIGEVLDGD	GTFVENHDNPR

Fig. 1-3: The four conserved regions and the corresponding β-sheets found in the amino acid sequence of α-amylase family 13 enzymes. Highlighted are the conserved catalytic amino acid residues. β2, β4, β5, and β7 indicate the β-sheet in which the region is present After(van der Maarel *et al.*, 2002)

1.4.1 Pullulan degrading enzymes

The pullulan (Fig. 1-3) degrading enzymes are classified into five groups on the base of substrate specificity and products formation. (1) pullulanase type I, (2) pullulanase type II, (3) pullulan hydrolase type I, (4) pullulan hydrolase type II, and (5) pullulan hydrolase type III.

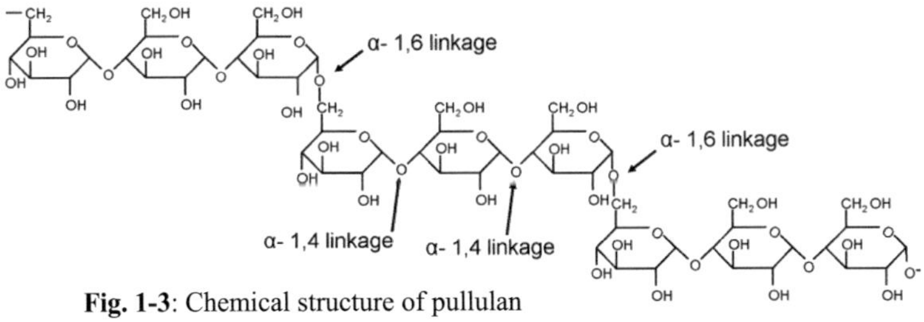

Fig. 1-3: Chemical structure of pullulan

1. *Pullulanase type I*: Pululanase type I hydrolyzes specifically α-1,6 linkages in pullulan (Fig. 1-3), amylopectin or limit dextrins with high specificity. Pullulan is completely degraded in a random fashion to maltotriose (Fig. 1-4). Substrates with short branches such as β-limit dextrin are hydrolyzed at a higher rate than amylopectin. Pullulanase type I requires at least two α-1,4 linked

glucose units in the vicinity of the α-1,6 linkages (Bertoldo & Antranikian, 2002a,b). Sequences information has revealed that there is low overall homology among pullulanase type I. However, a highly conserved region consisting of seven amino acids, Y/FNWGYDP, is found in all type I pullulanases described to date (Bertoldo *et al.*, 2004). Debranching enzymes, which exclusively hydrolyze the α-1,6 glycosyl bonds, are isoamylases (EC 3.2.1.68) and pullulanase type I (EC 3.2.1.41).

2. *Pullulanase type II or amylopullulanase*: Unlike pullulanase type I, pullulanase type II hydrolyzes α-1,6 linkages in pullulan and in addition, capable to cleave α-1,4 linkages in amylose. This enzyme with dual specificity belong to a new class of pullulanase, termed pullulanase type II or amylopullulanase. Pullulanase type I and II are absolutely unable to hydrolyze substrates like dextran or isomaltotriose (Fig. 1-4) which contain exlusivly α-1,6 linkages (Bertoldo & Antranikian, 2002a,b).

3. *Pullulan hydrolases (type I, II, and III)*: Pullulan hydrolase type I (neopullulanase) and pullulan hydrolase type II (isopullulanase) hydrolyze the α-1,4 linkages in pullulan, liberating panose and isopanose respectively (Fig. 1-4). Both enzymes are unable to hydrolyze α-1,6 linkages bonds in branched substrates or pullulan. According to this, the classification of pullulan hydrolyase into the group debranching enzymes is misleading. Recently, pullulan hydrolase type III has been detected. The enzyme attacks α-1,4 as well as α-1,6 linkages in pullulan, producing maltotriose, maltose, panose and glucose (Fig 1-4) (Bertoldo & Antranikian, 2002a,b).

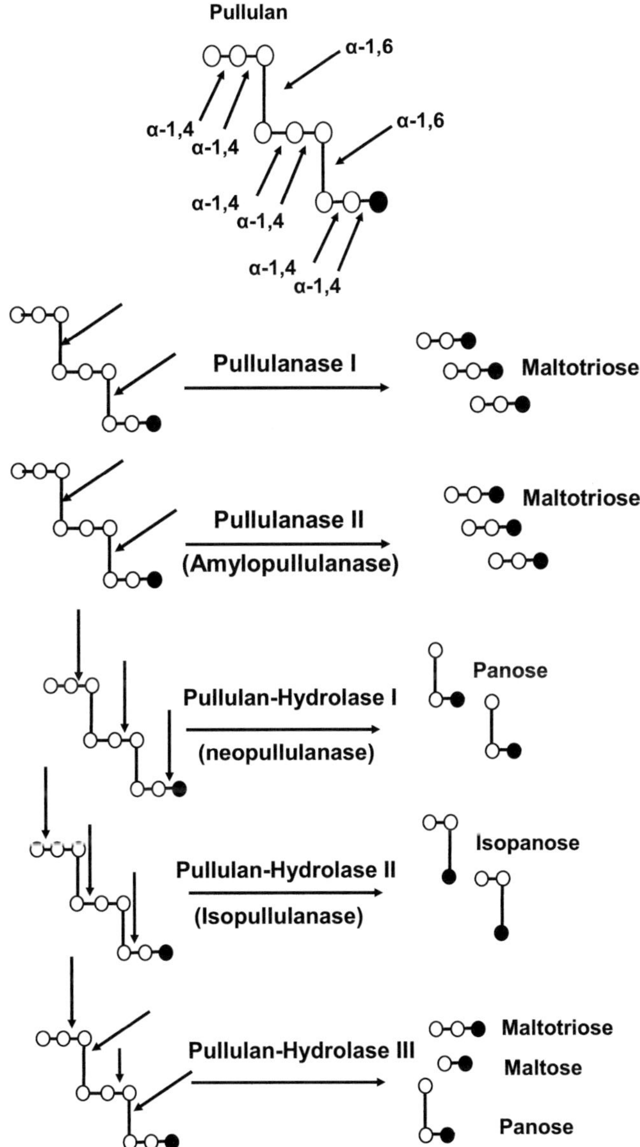

Fig. 1-4: The enzymatic action (arrows) of pullulanase type I and II, and pullulan hydrolase type I, II, and III. Black circles indicate reducing end.

1.4.2 Microbial sources and industrial application of pullulanase

True pullulanases have been found and isolated from a variety of bacteria. The pullulanase is synthesized intracellularly, then is usually secreted as an extracellular enzyme in the culture supernatant (MacGregor, 2003). Pullulanase type I is predominantly produced by mesophilic microorganisms, such as *Klebsiella pneumoniae, Bacillus acidopullulyticus, Bacillus flavocaldarius,* and *Bacillus deramificans* (Bertoldo & Antranikian, 2002a; Kelly *et al.*, 1994; Kornacker & Pugsley, 1990; MacGregor, 2003; Suzuki *et al.*, 1991). *Fervidobacterium pennivorans* is one of the few anaerobic bacteria which produce heat-stable pullulanase type I (Bertoldo *et al.*, 1999). Pullulanase type II is widely distributed in anaerobic microorganisims, inculding strains of the genera *Dictyoglomus, Thermoanaerobacter, Thermoanaerbium,* and *Pyrococcus*. The enzymes are extremely thermostable and are optimally active in the temperature range between 75 and 105 °C (Bertoldo & Antranikian, 2002a). The occurrence of pullulanase type I and II in microorganisms is listed in Table 1-4.

Microbial pullulanases are mainly used in food industry, specifically for the starch degradation. Pullulanase as a debranching enzyme cleaves α-1,6 linked branched points of starch and produce linear dextrins of varying chain length. Therefore, pullulanase is used in the production of glucose, fructose and maltose syrups (Vieille & Zeikus, 2001). However, before microbial pullulanase can be used in the production of glucose, fructose, and maltose syrups, starch in a slurry must be gelatinized by heat treatment and liquefied using a heat-stable α-amylase (Kujawski *et al.*, 2002; MacGregor, 2003). Type I pullulanases (hydrolyzing only α- 1,6 glucosidic linkages) are used in starch saccharification, leading to the production of glucose, fructose, and maltose syrups.

Type II pullulanase or amylopullulanase shows dual specificity by hydrolyzing α-1,4 and α-1,6 glucosidic linkages in starch. For this reason, they are not the best to be used for the maltose and glucose syrups production. Pullulanase type II, however, has been suggested as an alternative enzyme to replace α-amylases during starch liquefaction. Since certain pullulanase type II specifically produce maltose, maltotriose, and maltotetraose (DP2 to DP4) as the major end products of starch degradation, they have been suggested as catalysts in a one-step liquefaction-saccharification process for the production of high-DP2-to-DP4 syrups (Kujawski *et al.*, 2002; MacGregor, 2003).

Table.1-4: Occurrence of pullulanases type I and II in bacteria and archaea.

Enzyme	Microorganisms	Optimal enzyme properties		Reference
		Temp. (°C)	pH	
Pullulanase type I	*Bacillus acidopullulolyticus*	< 60	6	Kelly *et al.*, 1994
	Bacillus flavocaldarins	75	6.3	Suzuki *et al.*, 1991
	Thermus aquaticus			Plant *et al.*, 1986
	Thermus caldophilus	75	5.5	Kim *et al.*, 1996
	Bacillus thermoleovorans	75	5.6	Messaoud *et al.*, 2002
	Caldocellulosiruptor saccharolyticus			Albertson *et al.*, 1997
	Fervidobacterium pennivorans	80	6	Bertoldo *et al.*, 1999
	Thermotoga maritima	90	6	Bibel *et al.*, 1998
	Anaerobranca gottschalkii	70	8	Bertoldo *et al.*, 2004
	Klebsiella pneumoniae	60	5	Kornacker *et al.*, 1990
	Bacillus deramificans	< 60	6	MacGegor, 2003
Pullulanase type II	*Desulfurococcus mucosus*	85	5	Duffner *et al.*, 2000
	Pyrococcus furiosus	120	5.5-6.5	Dong *et al.*, 1997
	Pyrococcus woesei	100	6	Rudiger *et al.*, 1995
	Thermococcus hydrothermalis	117	5.5	Erra-Pujada *et al.*, 1999
	Thermococcus aggregans	100	6.5	Niehaus *et al.*, 2000
	Thermococcus celer	90	5.5	MacGegor, 2003
	Thermococcus litoratis	98	5.5	MacGegor, 2003
	Thermoanaerobacter ethanolicus	90	5.5	Bertoldo and Antranikian 2002a
	Thermotoga maritima	90	7.5	Bertoldo and Antranikian 2002a

Glucose, fructose, and maltose syrups are widely used in food. They are used in confectionary and hard confectionary, soft drinks, brewing and fermentation, jams, jellies, ice cream, conserve, sauces, caramel, baking, yoghurts, and canned fruits (Vieille & Zeikus, 2001). Pullulanases are also used to produce low carbohydrate (low calory) light beer and to produce branched cyclodextrins (CDs) which are more soluble in water than unbranched CDs (MacGregor, 2003).

1.5 Proteases

"Proteases are essential constituents of all forms of life on earth"
(Gupta *et al.*, 2002)

Proteases are the class of enzymes which occupy a pivotal position with respect to their applications in both physiological and commercial fields. Proteolytic enzymes catalyze the cleavage of peptide bonds in other proteins. Proteases are degradative enzymes which catalyze the total hydrolysis of proteins. Advances in analytical techniques have demonstrated that proteases conduct highly specific and selective modifications of proteins such as activation of zymogenic forms of enzymes by limited proteolysis, blood clotting and lysis of fibrin clots, and processing and transport of extracelular proteins across the membranes (Rao *et al.*, 1998).

Microbial proteases are among the most important hydrolytic enzymes and have been studied extensively since the advent of enzymology. The renewed interest in the study of proteases is mainly due to the recognition that these enzymes not only play an important role in the cellular metabolic processes, but have also gained considerable attention in the industrial community. Proteases have been widely used in industry, since their introduction in 1914 as detergent additives. The current estimated value of the worldwide sales of industrial protease is more than $1 billion (Gupta *et al.*, 2002; Rao *et al.*, 1998).

Microorganisms produce a large variety of proteases, which are intracellular and/or extracellular. Intracellular proteases are important for various cellular and metabolic processes such as: sporulation and differentiation, protein turnover, maturation of enzymes and hormones, and maintenance of the cellular protein pool. On the other hand, extracellular proteases are important for the hydrolysis of proteins in cell-free environments and enable the cell to absorb and utilize hydrolytic products (Kalisz, 1988). These extracellular proteases have also been commercially exploited to support protein degradation in various industrial processes (Kumar & Takagi, 1999)

1.5.1 Classification of proteases

According to the Nomenclature Committee of the International Union of Biochemistry and Molecular Biology, proteases are classified in subgroup 4 of group 3 (hydrolases). However, proteases do not comply easily with the general system of enzyme nomenclature due to their huge diversity of action and structure. Currently, proteases are classified on the basis of three major criteria: (1) type of reaction catalyzed, (2) chemical nature of the catalytic site, and (3) evolutionary relationship with reference to structure (Barett, 1994).

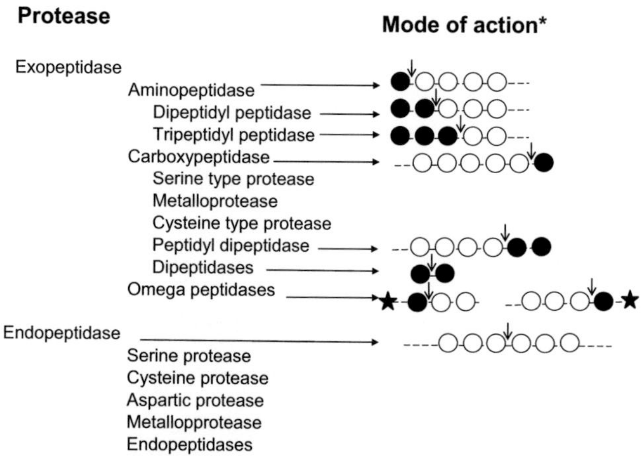

Fig. 1-5: Classification of proteases and mode of action. *Open circles represent the amino acid residues in the polypeptide chain. Solid circles indicate the terminal amino acids, and stars signify the blocked termini. Arrows show the sites of action of the enzyme (Nakajima *et al.*, 1986; Rao *et al.*, 1998)

Proteases are grossly subdivided into two major groups, i.e., exopeptidases and endopeptidases, depending on their site of action (Fig. 1-5). Exopeptidases cleave the peptide bond proximal to the amino or carboxy ends of the substrate, whereas endopeptidases cleave peptide bonds distant from the ends of the substrate. On the base of the functional group present at the active site, proteases are further classified into four prominent groups, i.e., serine proteases, aspartic proteases, cysteine proteases, and metalloproteases (Hartley, 1960). There are a few miscellaneous proteases that do not precisely fit into the standard classification, like., ATP-dependent proteases, which require ATP for activity (Menon & Goldberg, 1987). According to their amino acid sequences, proteases are classified into different families (Argos, 1987) and further subdivided into "clans" to accommodate sets of peptidases that have diverged from a common ancestor (Rawlings & Barrett, 1993).

1.5.1.1 Exopeptidases

The exopeptidases act only near the ends of polypeptide chains. Based on their site of action at the N or C terminus, they are classified as amino- and carboxypeptidases, respectively (Rao *et al.*, 1998). Aminopeptidases act at a free N-terminus of the polypeptide chain and liberate a single amino acid residue, a dipeptide, or a tripeptide (Fig. 1-5). Aminopeptidases occur in a wide variety of microorganisms including bacteria and fungi and in general, aminopeptidases are intracellular enzymes (Watson, 1976).

1.5.1.2 Endopeptidases

Endopeptidases are characterized by their preferential action at the peptide bonds in the inner regions of the polypeptide chain away from the N and C termini. The presence of the free amino or carboxyl group has a negative influence on enzyme activity. The endopeptidases are divided into four subgroups based on their catalytic mechanism, (1) serine proteases, (2) aspartic proteases, (3) cysteine proteases, and (4) metalloproteases. To facilitate quick and unambiguous reference to a particular family of peptidases, Rawlings and Barrett have assigned a code letter denoting the catalytic type, i.e., S, C, A, M, or U (see above) followed by an artibrarily assigned number (Rawlings & Barrett, 1993).

1. *Serine proteases*: Serine proteases are characterized by the presence of a serine group in their active site. Based on their structural similarities, serine proteases have been grouped into 20 families, which have been further subdivided into about six clans with common ancestors (Barett, 1994).

2. *Aspartic proteases:* Aspartic acid proteases, commonly known as acidic proteases, are the endopeptidases that depend on aspartic acid residues for their catalytic activity. Acidic proteases have been grouped into three families, namely, pepsin (A1), retropepsin (A2), and enzymes from pararetroviruses (A3). The members of families A1 and A2 are known to be related to each other, while those of family A3 show some relatedness to A1 and A2 (Rao *et al.*, 1998).

3. *Cysteine proteases*: Cysteine proteases occur in both prokaryotes and eukaryotes. About 20 families of cysteine proteases have been recognized. The activity of all cysteine proteases depends on a catalytic dyad consisting of cysteine and histidine. The order of Cys and His (Cys-His or His-Cys) residues differ among the families (Barett, 1994).

4. *Metalloproteases:* Metalloproteases are the most diverse of the catalytic types of proteases. They are characterized by the requirement for a divalent metal ion for their activity. About 30 families of metalloproteases have been recognized, of which 17 contain only endopeptidases, 12 contain only exopeptidases, and 1 (M3) contains both endo- and exopeptidases (Barett, 1995).

1.5.1.3 Serine proteases

Serine proteases, which are characterized by the presence of a serine group in their active site, are numerous and widespread among viruses, bacteria, and eukaryotes. Serine proteases are found in the exopeptidase, endopeptidase, oligopeptidase, and omega peptidase groups. According to their structural similarities, serine proteases have been grouped into 20 families, which have been further subdivided into about six clans (Fig. 1-6) with common ancestors (Barett, 1994). The primary structures of the members of four clans, chymotrypsin-like (SA), subtilisin-like (SB), carboxypeptidase C (SC), and *Escherichia* D-Ala–D-Ala peptidase A (SE) are totally unrelated, suggesting that there are at least four separate evolutionary origins for serine proteases. Clans SA, SB, and SC have a common reaction mechanism consisting of a common catalytic triad of the three amino acids, serine (nucleophile), aspartate (electrophile), and histidine (base). Although the geometric orientations of these residues are similar, the protein folds are quite different, forming a typical example of a convergent evolution. The catalytic mechanisms of clans SE and SF (repressor LexA) are distinctly different from those of clans SA, SB, and SE, because they lack the classical Ser-His-Asp triad. Another interesting feature of the serine proteases is the conservation of glycine residues in the vicinity of the catalytic serine residue to form the motif Gly-Xaa-Ser-Yaa-Gly (Brenner, 1988).

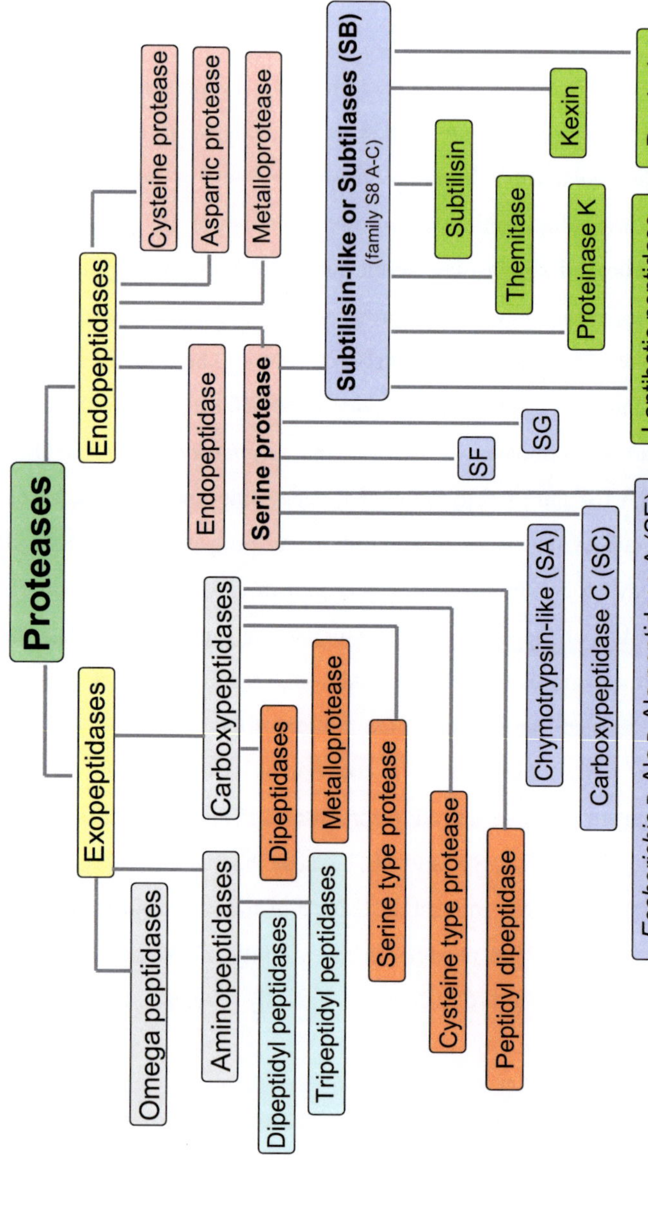

Fig. 1-6: Classification of proteases (Gödde *et al.*, 2005; Gupta *et al.*, 2002; Rao *et al.*, 1998; Siezen and Leunissen, 1997)

Serine proteases are irreversibly inhibited by 3,4 dichloroisocoumarin (3,4-DCI), L-3-carboxytrans 2,3-epoxypropyl-leucylamido (4-guanidine) butane (E.64), diisopropylfluorophosphate (DFP), phenylmethylsulfonyl fluoride (PMSF) and tosyl-L-lysine chloromethyl ketone (TLCK). Thiol reagents such as *p*-chloromercuribenzoate (PCMB) and dithiothreitol (DTT) inhibit the serine proteases that have a cysteine residue near the active site (Rao *et al.*, 1998).

1.5.1.3.1 Subtilisin-like or Subtilases (SB)

Serine endo- and exo-peptidases are of extremely widespread occurrence and diverse function. They are distinct families of serine proteases and they have been grouped into six clans, of which the two largest are the chymotrypsin-like and subtilisin-like clans (Barett, 1995; Rawlings & Barrett, 1994). The superfamily (Clan) Subtilisin-like (SB) (S8 A-C) has been grouped into six familys: Subtilisin, Thermidase, Kexin, Proteinase K, Pyrolysin, and Lantibotic peptidase (Fig. 1-6) (Gödde *et al.*, 2005; Siezen & Leunissen, 1997).

1. *Subtilisin*: Only found in microorganisms. Numerous miner variants of true subtilisin have been identified such as, subtilisin carlsberg and subtilisin BPN (Gödde *et al.*, 2005; Siezen & Leunissen, 1997).

2. *Thermitase*: Found in microorganisms, including some thermophiles and halophiles. Aerolysin and fervidolysin are two examples of thermitase and recently a new thermitase was identified from *Fervidobacterium islandicum* as islandisin (Gödde *et al.*, 2005).

3. *Proteinase K*: It is a large family of endopeptidases found only in fungi, yeast and gram-negative bacteria. Aqualysin is an examples of proteinase K (Siezen & Leunissen, 1997).

4. *Pyrolysin*: Heterogeneous group of enzymes of different origin. Stettelysin is an examples of pyrolysin (Siezen & Leunissen, 1997).

5. *Lantibiotic peptidase*: A small number of highly specialized enzymes for cleavage of leader peptides from precursors of lantibiotics, a unique group of post-translationally modified antimicrobial peptides. These endopeptidases have only been found in gram positive bacteria and several are intracellular (Siezen & Leunissen, 1997).

6. *Kexin*: A large group of proprotein convertases (PCs) have been identified and all of them are involved in activation of peptide hormones, growth factors, viral proteins, etc.. High specificity has been observed for cleavage after dibasic (Lys-Arg or Arg-Arg) or multiple basic residues, and nearly all are eukaryotic (Barr, 1991).

1.5.2 Microbial sources of proteases

Proteases are essential constituents of all forms of life on earth, including prokaryotes, fungi, plants and animals. They can be produced in large quantities in a relatively short time by established methods of fermentation. Microorganisms account for a two-third share of commercial protease production in the world (Kumar & Takagi, 1999). Microbial proteases are classified into various groups, dependent on whether they are active under acidic, neutral, or alkaline conditions and

on the characteristics of the active site group of the enzyme, i.e. metallo-
(EC.3.4.24), aspartic- (EC.3.4.23), cysteine- or sulphydryl- (EC.3.4.22),
or serine-type (EC.3.4.21) (Kalisz, 1988). Some microbial sources of
serine subtilisin-like (subtilase) proteases are listed in Table 1-5.

Table.1-5: Formation of subtilase proteases in bacteria and eukarya

Organism	Enzyme	References
BACTERIA		
Bacillus subtilis A50	Intracell. serine protease	Strongin *et al.*, 1978
Bacillus sp. Gx6644	Subtilisin GX	Durham, 1993
Bacillus sp. Y	Protease BYA	Shimogaki *et al.*, 1991
Bacillus thuringiensis israelensis	Extracellular serine protease	Chestukhina *et al.*, 1980
Bacillus thuringiensis finitimus	Extracellular serine protease	Chestukhina *et al.*, 1980
Bacillus thuringiensis kurstaki	Extracellular serine protease	Kunitate *et al.*, 1989
BaciNus cereus	Extracellular serine protease	Chestukhina *et al.*, 1980
Thermus Tok3A 1	Caldolysin	Freeman *et al.*, 1993
Vibrio metschnikovii	Alkaline protease VapK	Kwon *et al.*, 1995
EUKARYA		
Fungi		
Agaricus bisporus	Extracellular serine protease	Burton *et al.*, 1993
Malbranchea suljurea	Thermomycolin	Gaucher & Stevenson, 1976
Ophiostoma piceae	Extracellular protease	Abraham & Breuil, 1995
Verticillium chlamydosporium	Extracellular protease	Segers *et al.*, 1995
Scedosporium apiospermum	Extracellular protease	Larcher *et al.*, 1996

Most commercial proteases, mainly neutral and alkaline, are produced
by micoorganisms belonging to the genus *Bacillus*. Bacterial neutral
proteases are active in a narrow pH range (pH 5 to 8) and have relatively
low thermostability. Due to the intermediate rate of reaction, neutral
proteases generate less bitterness in hydrolyzed food proteins than the
animal proteinases and hence are valuable for use in the food industry.
Neutrase, a neutral protease, is insensitive to the natural plant proteinase
inhibitors and is therefore useful in the brewing industry (Rao *et al.*,
1998). Bacterial alkaline proteases are characterized by their high
activity at alkaline pH (pH 10), their broad substrate specificity, and
optimal activity temperature around 60°C. These properties of bacterial
alkaline proteases make them suitable for the use in the detergent
industry (Gupta *et al.*, 2002; Rao *et al.*, 1998).

Fungi produce a wider variety of enzymes than bacteria. For example, in *Aspergillus oryzae* acid, neutral, and alkaline proteases have been detected. The fungal proteases are active in wide pH range (pH 4 to 11) and exhibit broad substrate specificity. They have, however, a lower reaction rate and they are less thermostable than the bacterial enzymes. Fungal enzymes can be conveniently produced in a solid-state fermentation process. Fungal acid proteases have an optimal pH between 4 and 4.5 and are stable between pH 2.5 and 6.0 (Rao *et al.*, 1998). Some microbial sources of serine subtilisin-like (subtilase) proteases and their properties are listed in Table 1-6

Table.1-6: Subtilase proteases from bacteria and archaea. +: enzyme inhibited.

Microorganisms	Enzyme Properties					Reference
	Optimal:		Inhibited by:			
	Temp. (°C)	pH	Antipain	Chymostatin	PMSF	
BACTERIA						
Pseudoalteromonas sp.	30	8.8	+	+	+	Lee *et al.*, 2000
Shewanella strain AC10	20	9.0				Kulakova *et al.*, 1999
Streptomyces albogriseolus	70-80	10.0	+	+	+	Suzuki *et al.*, 1997
Mycobacterium tuberculosis				+	+	Dave *et al.*, 2002
Xenorhabdus nematophila	23	7.0	+	+		Caldas *et al.*, 2002
Thermoanaerobacter yonseiensis	92	9.0			+	Jang *et al.*, 2002
Bacillus halmapalus	60-70	11.0-12.0				Saeki *et al.*, 2002
Fervidobacterium pennivorans				+	+	Kluskens *et al.*, 2002
Fervidobacterium islandicum	80	8.0			+	Gödde *et al.*, 2005
ARCHAEA						
Thermococcus kodakaraensis	80	9.5				Kannan *et al.*, 2001
Aeropyrum pernix K1	90	8.0			+	Catara *et al.*, 2003
Natrialba magdii	60	8.0-10.0		+	+	Gimenez *et al.*, 2000
FUNGI						
Pleurotus ostreatus	37	4.5	+	+	+	Palmieri *et al.*, 2001

1.5.3 Industrial applications of proteases

Proteases have a large variety of applications but they are mainly used in the detergent and food industries. In view of the recent trend of developing environmentally friendly technologies, proteases are envisaged to have extensive applications in leather treatment and in several bioremediation processes. Proteases are used extensively in the pharmaceutical industry for preparation of medicines such as ointments for debridement of wounds, etc. Proteases that are used in the food and detergent industries are prepared in bulk quantities and used as crude preparations, whereas those that are used in medicine are produced in small amounts but require extensive purification before they can be used (Gupta *et al.*, 2002; Rao *et al.*, 1998). Some commercial bacterial alkaline proteases, soures, applications and their industrial suppliers are listed in Table 1-7.

In summary the wide specificity of the hydrolytic action of proteases finds an extensive application in the industry and the most important examples are listed below (Gupta *et al.*, 2002; Rao *et al.*, 1998):

1. *Food industry*: The use of proteases in the food industry dates back to antiquity. They have been routinely used for various purposes such as cheese making, baking, preparation of soya hydrolysates, and meat tenderization.
2. *Detergents*: Proteases are one of the standard ingredients of all kinds of detergents ranging from those used for household laundering to reagents used for cleaning contact lenses or dentures.

3. *Leather industry*: Leather processing involves several steps such as soaking, dehairing, bating, and tanning. The major building blocks of skin and hair are proteinaceous. Proteases are used for selective hydrolysis of noncollagenous constituents of the skin and for removal of nonfibrillar proteins such as albumins and globulins.

4. *Pharmaceutical industry*: The wide diversity and specificity of proteases are used to great advantage in developing effective therapeutic agents.

5. *Research*: proteases play an important role in basic research. The selective peptide bond cleavage is used in the elucidation of structure function relationship, in the synthesis of peptides and in the sequencing of proteins.

Table.1-7: Commercial bacterial alkaline proteases, soures, applications and their industrial suppliers.

Supplier	Product trade name	Microbial source	Application
Novo Nordisk, Denmark	Alcalase	*Bacillus licheniformis*	Detergent, silk degumming
	Savinase	*Bacillus sp.*	Detergent, textile
	Esperase	*B. lentus*	Detergent, food, silk degumming
	Biofeed pro	*B. licheniformis*	Feed
	Durazym	*Bacillus sp.*	Detergent
	Novozyme 471MP	n.s.	Photographic gelatin hydrolysis
	Novozyme 243	*B. licheniformis*	Denture cleaners
	Nue	*Bacillus sp.*	Leather
Genencor International, USA	Purafact	*B. lentus*	Detergent
	Primatan	Bacterial source	Leather
Gist-Brocades, The Netherlands	Subtilisin	*B. alcalophilus*	Detergent
	Maxacal	*Bacillus sp.*	Detergent
	Maxatase	*Bacillus sp.*	Detergent
Solvay Enzymes, Germany	Opticlean	*B. alcalophilus*	Detergent
	Optimase	*B. licheniformis*	Detergent
	Maxapem	Protein engineered variant of Bacillus sp.	Detergent
	HT-proteolytic	*B. subtilis*	Alcohol, baking, brewing, feed, food, leather, photographic waste
	Protease	*B. licheniformis*	Food, waste
Amano Pharmaceticals, Japan	Proleather	*Bacillus sp.*	Food
	Collagenase	*Clostridium sp.*	Technical
	Amano protease S	*Bacillus sp.*	Food
Enzyme Development, USA	Enzeco alkaline protease	*B. licheniformis*	Industrial
	Enzeco alkaline protease	*B. licheniformis*	Food
	Enzeco high alkaline protease	*Bacillus sp.*	Industrial
Nagase Biochemicals, Japan	Bioprase concentrate	*B. subtilis*	Cosmetic, pharmaceuticals
	Ps. protease	*Pseudomonas aeruginosa*	Research
	Ps. elastase	*Pseudomonas aeruginosa*	Research
	Cryst. protease	*B. subtilis (K2)*	Research
	Cryst. protease	*B. subtilis (bioteus)*	Research
	Bioprase	*B. subtilis*	Detergent, cleaning
	Bioprase SP-10	*B. subtilis*	Food
Godo Shusei, Japan	Godo-Bap	*B. licheniformis*	Detergent, food
Rohm, Germany	Corolase 7089	*B. subtilis*	Food
Wuxi Synder Bioproducts, China	Wuxi	*Bacillus sp.*	Detergent
Advance Biochemicals, India	Protosol	*Bacillus sp.*	Detergent

Table after Gupta *et al.*, 2002

1.6 The aim of the study

Microbial life does not seem to be limited to specific environments. During the past few decades it has become clear that microbial communities can be found in the most diverse conditions, including extremes of temperature, pressure, salinity and pH. These microorganisms, called extremophiles, produce biocatalysts that are functional under extreme conditions. Consequently, the unique properties of these biocatalysts have resulted in several novel applications of enzymes in industrial processes. At present, only a minor fraction of the microorganisms on Earth have been exploited. Novel developments in the cultivation and production of extremophiles, as well as also developments related to the cloning and expression of their genes in heterologous hosts, will increase the number of enzyme-driven transformations in chemical, food, pharmaceutical and other industrial applications. For this reason, the aims of this study are:

1. Isolation and characterization of novel psychrophilic bacteria from seawater samples obtained from the area of Spitsbergen and investigating their ability to produce cold-active enzymes.

2. Cloning and characterization of novel cold-adapted enzymes and investigating their potential to be used in industry

2 MATERIALS & METHODS

2.1 Isolation and characterization of aerobic and anaerobic bacteria.

2.1.1 Sample collection, media and culturing conditions

Arctic sea ice and seawater samples were collected in 1998 during an expedition in Spitsbergen, Norway. Samples were transported to the laboratory at an ambient temperature of 10 °C and 1 ml volume of liquid sample from 50 samples was incubated in a complex marine anaerobic liquid medium under anaerobic conditions and in a complex marine aerobic liquid medium.

The complex anaerobic medium contained (values in g l^{-1}): NaCl 28.13 g; KCl 0.77 g; $CaCl_2$ x $2H_2O$ 0.02 g; $MgSO_4$ x$7H_2O$ 0.5 g; NH_4Cl 1 g; 10-fold concentration Trace element solution (DSM 141) 1 ml; 10-fold concentration vitamin solution (DSM 141) 1 ml; KNO_3 1 g; KH_2PO_4 x$2H_2O$ 0.5 g; yeast extract 2.5 g; peptone 2.5 g; Na-Acetate 0.1 g; Na_2-Succinate 0.1 g; DL-Malate 0.1 g; Na-Pyruvate 0.1 g; Glucose 1 g; Cysteine 0.3 g; Rezazurin (1 % solution) 100 µl. The medium without cysteine was boiled, concentrated to 90 % of the original volume and cooled to 4 °C under continuous gassing with N_2. Cysteine was added to distribution in 15 ml Hungate tubes (9 ml medium per tube) or 100 ml serum bottles (20 or 50 ml per bottle); the final pH was adjusted to 7. For the isolation of a pure culture, serial dilution technique was applied. The growth was carried out at 4-10 °C for 1 -2 weeks and was measured by determining the optical density at 600 nm (1–cm path length) using Schimadzu UV spectrophotometer.

The complex marine aerobic medium consisted of a basal medium supplemented with a solution of different carbon sources. The basal medium contained (volume g l^{-1}): NaCl 28.13 g; KCl 0.77 g; $CaCl_2$ x2H$_2$O 0.02 g; $MgSO_4$ x7H$_2$O 0.5 g; NH_4Cl 1.0 g; iron-ammonium-citrate 0.02 g; yeast extract 0.5 g; 10-fold concentration trace element solution (DSM 141) 1 ml; 10-fold concentration vitamin solution (DSM 141) 1 ml; KH_2PO_4 2.3 g; Na_2HPO_4 x2H$_2$O 2.9 g. The carbon source mixture solution (volume g l^{-1}): Na-acetate 0.5 g; Na_2-succinate 0.5 g; Na-pyruvate 0.5 g; DL-malate 0.5 g; D-mannitol 0.5 g and glucose 2 g. The final pH of the complex medium was 7. Incubation was carried out at 4 °C for about 4-7 days, before growth and colonies on agar plates were observed. Colonies were selected on the basis of morphological differences. For the isolation of a pure culture, serial dilution and plating techniques were applied. The pure isolates were routinely cultivated on complex marine medium agar plates at 15 °C for 4 days.

2.1.2 Cellular characterization

Gram staining test was performed by staining cells using the Hucker method (Gerhardt *et al.*, 1994). For the sporulation test, cells were grown for up to 6 days in medium containing no carbon source but with

0.1 % (w/v) yeast extract. The present of spores and the cell morphology was determined by phase-contrast microscope (Zeiss/Axioskop).

2.1.3 Optimal temperature, pH and salt requirement for growth

For the anaerobic strain: The optimum growth temperature was tested between 4-50 °C and pH 7. The pH optimum for growth was tested between pH 2-9 at 10 °C. The salt requirement was determined at different NaCl concentrations between 0 and 6 % (w/v), with no change of the other salts concentrations, at pH 7 and 10 °C.

For the aerobic strain: The optimum growth temperature was tested between 4-37 °C and pH 7. The pH optimum for growth was tested between pH 2-10 at 15 °C. The salt requirement was determined at different NaCl concentrations between 0 and 10 % (w/v), with no change of the other salts concentrations at pH 7 and 15 °C.

Growth for both strains was measured by determining the optical density at 600 nm (1–cm path length) using Schimadzu UV spectrophotometer.

2.1.4 Substrate spectrum

The growth of the isolated anaerobic strain was tested on different substrates as the sole carbon source by sub-culturing the new strain three times into a medium containing 10 mM of the following substrates:

succinate, pyruvate, glucose, starch, fumarate, citrate, fructose, maltose, malate, arabinose, crotonate, oxaloacelate, L-asparatate, lysine, threonine, glutamate, cysteine, L-tartate, D-tartate, 3 hydroxybutyrate, glycerol, xylose or yeast extract. Incubation was carried out at 10 °C for 4 days. Growth was measured by determining the optical density of the cells at 600 nm (1–cm path length) using Schimadzu UV spectrophotometer.

Substrates utilization from the isolated aerobic strain was tested on api 20 NE strips (20050) (Bio Merieux.Inc), api 20 E strips (20 100/20 160) (Bio Merieux.Inc) and Biolog GN2 micro-plates. Api 20 NE strips (20050) (Bio Merieux.Inc) and api 20 E strips (20 100/20 160) were also used for testing the ability of the new strain to produce indol (tryptophane), acetoin and H_2S (from sodium thiosulfate), and the ability of the strain to reduce nitrate to nitrite and nitrite to nitrogen. The strips and micro-plates were inoculated with the new strain cells suspension (cells were resuspended in NaCl 0.85 % (w/v) medium to a final OD_{600nm} 0.5) and incubated at 15 °C overnight.

A large number of carbon source were tested: α cyclo-dextrin, dextrin, Tween 80, Tween 40, N-acetyl-D-glucosamine, α-D-glucose, maltose, sucrose, methylpyruvate, D,L-lactic acid, succinic acid, bromo succinic acid, inosine, esculin ferric citrate, L-arabinose, potassium gluconate, malic acid, trisodium citrate, glycogen, N-acetyl-D-galacto-samine, adonitol, capric acid, D-arabitol, cellobiose, L-erythol, D-fructose, L-fructose, D-galactose, gentiobiose, m-inositol, α-D-lactose, lactulose, D-mannitol, D-mannose, D-mellobiose, β-methyl-D-glucose, D-psicose, D-raffinose, L-rhamnose, D-sorbitol, D-trehalose, turanose, Xylitol, mono-methyl succinate, acetic acid, cis-aconitic acid, citric acid, formic acid, D-galactonic acid lactone, D-galacturonic acid, D-gluconic acid, D-glucosaminic acid, D-glucuronic acid, α-hydroxy-butyric acid, β-hydroxy-butyric acid, γ-hydroxy-butyric acid, p-hydroxyphenylacetic acid, itaconic acid, α-keto butyric acid, α-keto glutaricacid, α-ketovaleric acid, malonic acid, propionic acid, quinic acid, D-saccharic acid, seabacic acid, succinamic acid, glucuronamide, alaninamide, D-alanine, L-alanine, L-alanyl-glycine, L-asparagine, L-aspartic acid, L-glutamic acid, glycyl-L-aspartic acid, glycyl-L-glutamic acid, L-histidine, hydroxy L-proline, L-leucine, L-ornithine, L-phenyl-alanine, L-proline, L-pyroglutamic acid, D-serine, L-serine, L-threonine, D,L-carnitine, γ-amino butyric acid, urocanic acid, uridine, thymidine, phenylthylamine, putrescine, 2-aminoethanol, 2,3-butanediol, glycerol, D,L α-glycerol phosphate, glucose-1-phosphate, glucose-6-phosphate or phenylacetic acid.

2.1.5 Fermentation products

Serum bottles (50 ml per bottle) were prepared with the same anaerobic medium described in section 2.1.1. In this medium however, only one substrate was used at a time as a carbon source (5 g l^{-1}); starch, glucose or pyruvate. The isolated anaerobic strain was cultured in the different carbon source media and incubated for 4 days at 10 °C.

Lactate, formate, ethanol, acetate, propionate, iso-butyrate and n-butyrate were analyzed as fermentation products. The analysis of formate was performed by HPLC- UV- detection. Headspace–

Gaschromotograph HS-GC, HS 100 Perkin Elmer, Sigma 2000 Perkin Elmer capillary column, FID was used to detect acetate, ethanol, propionate, iso-butyrate and n-butyrate.

2.1.6 Electron acceptors

Serum bottles (50 ml per bottle) were prepared with sulfate and nitrate - free anaerobic medium containing (values in g l^{-1}): NaCl 28.13 g; KCl 0.77 g; $CaCl_2$ $x2H_2O$ 0.02 g; NH_4Cl 0.5 g; 10 fold concentration vitamin solution (DSM 141) 1 ml; KH_2PO_4 $x2H_2O$ 0.6 g; yeast extract 0.1 g; glucose 5 g; Cysteine 1 g; Resazurin (1% solution) 100 µl. The anaerobic medium preparation was carried out as described in section 2.1.1. To the medium one of the following electron acceptors was added: sulfate (10 mM), sulfur (0.1 % (w/v)), thiosulfate (10 mM) or nitrate (10 mM). The isolated anaerobic strain was cultured in the different media and incubated for 4 days at 10 °C. The sulfate, thiosulfate, sulfur and nitrate reduction was tested by sulfate and nitrate analytical test strips (Mekquant 1.10019.00 and 1.14773.001 respectively) produced by Merck, Germany. The fermentation products were analyzed in the same way as descried in section 2.1.5.

2.1.7 Enzyme production

The ability of the isolated aerobic strain to produce alkaline phospatase, arginine dihydrolase, esterase (C4), esterase lipase (C8), lipase, leucine arylamidase, valine arylamidase, cystine arylamidase, trypsin, α chymotrpsin, acid phosphatase, omithine decarboxylase, lysine decarboxylase, urease, tryptophane deaminase, naphthol-AS-BI-phosphohydrolase, α and β–galactosidase, β–glucoronidase, α- and β–glucosidase, N-acetyl-β–glucosaminidase, α–mannosidase or α–fucosidase was tested on api zym (25 200) strips (bio Merieux, Inc) by inoculation of the strips wells with cell suspension (cells were resuspended in water to a final OD_{600nm} 5.0) of the isolated aerobic strain. The strips were incubated at 15 °C for 6 hours.

Amylase, arabinase, arabinoxylanase, protease, HE-cellulase, glucanase, dextranase, galactanase, galactomannanase, β-glucanase, pullulanase, curdlanase, xylanase and xyloglucanase production by the isolated aerobic strain was tested on diffusion agar plates containing the base medium and 0.1 % (w/v) of one of the following substrates: red pullulan (pullulanase), azo-casein (protease) , AZCL–pullulan (pullulanase), AZCL–HE–cellulose (HE–celulase), AZCL–arabinan (arabinase), AZCL–arabinoxylan (arabinoxylanase), AZCL–curdlan (curdlanase), AZCL–amylose (amylase), AZCL–dextran (dextranase), AZCL–galactan (galactanase), AZCL–galactomannan (galactomannanase), AZCL–β-glucan (β–glucanase) AZCL–xylan (xylanase) and AZCL–xyloglucan (xyloglucanase). Substrate degradation was detected by clearing zone/color diffusion halo around the colonies after the isolated aerobic stain was grown on these substrates agar plates at 10 °C for 2-4 days. AZCL – polymers were purchased from Megazyme, Bray, Ireland.

2.1.8 Fatty acids analysis

Cells of the isolated strains were harvested from 2 l culture sample by centrifugation and were used for fatty acids analysis. Fatty acids methyl ester (FAME) analysis was performed according to the modified method of (Lepage & Roy, 1984) Lepage & Roy (1984). Total lipids were extracted according to (Bligh & Dyer, 1959) Bligh & Dyer (1959). The FAMEs were analyzed by capillary gas chrompack (Kohn *et al.*, 1996). A fused silica capillary column D23, 40 m (Fisons) was used for the separation of fatty acid species.

The chromatographic conditions were as follows: injector (PTV): 65 °C–270 °C split ratio 15:1; carrier gas: helium at a 40 cm s^{-1} flow. Column oven: initial temperature : 60 °C for 0.1 min; from 60 °C to 180 °C at 40 °C min^{-1}; 180 °C for 2 min; from 180 °C to 210 °C for 3 min; from 210 °C to 240 °C at 3 °C min^{-1}; 240 °C for 10 min.

2.1.9 16S rDNA amplification and sequencing

Cells of the isolated strains were harvested from a 500 µl culture sample by centrifugation and resuspended in 100 µl of water. A sample of 1 µl was used as a template for the amplification of the 16S rDNA. PCR-mediated amplification of the 16S rDNA was carried out according to Rainey & Stackebrandt (Rainey & Stackebrandt, 1993). PCR products were purified using the QIA quick PCR purification kit (Qiagen). Purified PCR products were directly sequenced using the Tag Dye Deoxy Terminator Cycle sequencing kit (Applied Biosystems). PCR product reactions were electrophoreted using Applied Biosystems model

373S DNA sequencer. Both strands of amplification products were sequenced using primers 8F, 518F and 1504R (Buchholz-Cleven *et al.*, 1997). The complete 16S rDNA sequence of the isolated strains was determined by the assembly of all sequence products using Pregap 4 version 1.4bl and Gap 4 version 4.8bl software.

2.1.10   Phylogenetic analysis

BLAST analysis was performed by NCBI online database of the 16S rDNA sequence to determine the phylogenetic grouping to which the isolated strains were most closely related. Reference sequences utilized in phylogenetic analysis were retrieved from NCBI database and aligned with the newly determined sequence of the isolated strains by using CLUSTAL W (1.83) software. The phylogenetic and molecular evolutionary analyses was performed with the neighbour joining method by using software from PHYLIP, version 3.57c (Felsenstien, 1995). The DNADIST program with Kimura-2 factor was used to compute the pairwise evolutionary distances for the above aligned sequences (Kimura, 1980). The topology of the phylogenetic trees was evaluated by performing a bootstrap (algorithm version 3.6 b) with 1000 bootstrapped trials. The phylogenetic trees were drawn using Tree View 32 software. As the out group the 16S rDNA of *Bacillus subtilis* was used.

The 16S rDNA sequence data of the isolated anaerobic strain was compared with all currently available sequences of organisms belonging to the *Fusobacteriaceae* family which are: *Ilyobacter tartaricus*

(AJ307982.1), *Ilyobacter insuetus* (AJ307980.1), *Ilyobacter polytropus* (AJ307981) *Propionigenium maris* (Y16800.1), *Propionigenium modestum* (X54275.1), *Fusobacterium necrophorum* (AF044948.1), *Fusobacterium equorum* (AJ295750.1) *Fusobacterium nucleatum* (AF481217.1), *Fusobacterium canifelinum* (AY162222.1), *Fusobacterium genomosp* (AY278616.1), *Fusobacterium genomosp* (AY278617.1), *Fusobacterium naviforme* (AJ006965.1), *Fusobacterium simiae* (AF342838.1), *Leptotrichia buccalis* (X90831.1), *Leptotrichia genomosp* (AY278621.1), *Leptotrichia goodfellowii* (AY029807.1), *Leptotrichia hofstadii* (AY029803.1), *Leptotrichia shahii* (AY029806.1), *Leptotrichia tervisanii* (AY029805.1), *Leptotrichia wadeii* (AY029802.1), *Sebaldella termitidis* (M58678.1), *Leptotrichia sanguinegens* (L37789.1) and *Streptobacillus moniliformis* (Z35305.1).

The 16S rDNA sequence data of the isolated aerobic strain was compared to all currently available sequences of the micoorganisms belonging to the genus *Shewanella*: *Shewanella putrefaciens* (U91552.1), *Shewanella affinis* (AF500080.1), *Shewanella alga* (U91544.1) *Shewanella denitrificans* (AJ457093.1), *Shewanella amazonensis* (AF005248.1), *Shewanella aquimarina* (AY485225.1), *Shewanella baltica* (AJ000214.1), *Shewanella benthica* (X82131.1) *Shewanella colwelliana* (AY653177.1), *Shewanella decolorationis* (AJ609571.1), *Shewanella fidelia* (AF420313.1), *Shewanella frigidimarina* (AJ300833.1), *Shewanella gaetbuli* (AY190533.1) *Shewanella gelidimarina* (U85907.1), *Shewanella hanedai* (X82132.1), *Shewanella japonica* (AF500079.1), *Shewanella kaireiae* (AB094598.1), *Shewanella livingstonis* (AJ300834.1), *Shewanella marinintestina* (AB081759.1), *Shewanella marisflavi* (AY485224.1), *Shewanella massilia* (AJ006084.1) *Shewanella pacifica* (AY366086.1),

Shewanella pealeana (AF011335.1), *Shewanella saccharophilus* (AF033028.1), *Shewanella sairae* (AB081762.1), *Shewanella schlegeliana* (AB081761.1), *Shewanella surugaensis* (AB094597.1), *Shewanella violacea* (D21225.1), *Shewanella waksmanii* (AY170366.1), *Shewanella woodyi* (AF003548.1) and *Shewanella olleyana* (AF295592.1).

2.1.11   G + C content of genomic DNA

Cells of the isolated strains were harvested from a 2 l culture sample by centrifugation and were used for the determination of G + C content of the genomic DNA. The cells were disrupted with a French press and DNA was purified by chromatography on hydroxyapatite (Cashion *et al.*, 1977). The mol % G + C of genomic DNA was determined by high performance liquid chromatography (HPLC) (Shimadzu corp, Japan) (Mesbah & Whitman, 1989). The analytical column used was a VYDAC 201 SP 54, C18 5 μm (250 x 4.6 mm) equipped with a guard column 201 GD 54H (Vydac, Hesperia, USA).

2.1.12   DNA – DNA hybridization

DNA–DNA hybridization of the isolated aerobic strain with the strain *Shewanella putrefaciens* (DSM 6067 = ATCC 8071) was carried out from 3 g cell mass of each strain. DNA was isolated using French press (Thermo spectroic) and was purified by chromatography on hydroxyapatite as described by (Cashion *et al.*, 1977). DNA–DNA hybridization was carried out as described by (De Ley *et al.*, 1970), with the modification described by (Huss *et al.*, 1983), using a model 2600 spectrophotometer equipped with a model 2527-R thermoprogrammer and plotter (Gilford Instrument Laboratories). Renaturation rates were computed with the TRANSFER. BAS program of (Jahnke, 1992).

2.2 Cloning, sequencing and expression of type I pullulanase and Subtilisin-like serine protease from *Shewanella arctica*. sp. nov.

2.2.1 Bacterial strain, medium and growth condition

Shewanella arctica sp. nov. was grown aerobically in a complex marine medium consisting of basal medium supplemented with a solution of different carbon sources. The basal medium contained (volume gl^{-1}): NaCl 28.13 g; KCl 0.77 g; $CaCl_2$ $x2H_2O$ 0.02 g; $MgSO_4$ $x7H_2O$ 0.5 g; NH_4Cl 1.0 g; iron-ammonium-citrate 0.02 g; yeast extract 0.5 g; 10-fold concentration trace element solution (DSM 141) 1 ml; 10-fold concentration vitamin solution (DSM 141) 1 ml; KH_2PO_4 2.3 g; Na_2HPO_4 $x2H_2O$ 2.9 g. The carbon source mixture solution (volume $g l^{-1}$): pullulan 10 g; Na-acetate 0.5 g; Na_2-succinate 0.5 g; Na-pyruvate 0.5 g; DL-malate 0.5 g; D-mannitol 0.5 g and glucose 2 g. The final pH of the complex medium was adjusted to 7. Incubation was carried out at 15 °C for 4 days.

2.2.2 Construction of a *Shewanella arctica* sp. nov. λ-phage gene library

The lambda ZAP Express Predigested vector and ZAP Express Predigested Gigapack cloning kits (*Bam*H1/ CIAP-Treated) (Stratagene) were used for construction a *Shewanella arctica* sp. nov. λ-phage gene library. *E. coli* XL1-Blue MRF' and *E. coli* XLOLR were used as host cells as described by the supplier (Stratagene). Chromosomal DNA of *Shewanella arctica* sp. nov. was isolated by utilizing Qiagen's genomic DNA isolation kit (Qiagen GmbH, Hilden). The genomic DNA was partially digested by *Sau*3A (New England, Biolabs, Inc.). Fragments were separated by agarose gel electrophoresis and 6–10 kb fragments were isolated by electroelution and phenol / chloroform extraction.

Shewanella arctica sp. nov. DNA fragments were ligated into the λ-ZAP express vector containing the phagemid pBK-CMV. After packaging of the ligation products the primary phage library was amplified in *E. coli* XL1-Blue MRF'. After amplification phagemid pBK-CMV harboring the insert DNA was excised using helper phage ExAssist. The excised ExAssist population was transfected and stably established as plasmid in *E. coli* XLOLR cells.

2.2.3 Screening for recombinant pullulanase and protease

The pBK-CMV plasmid, in *E. coli* XLOLR, contains IPTG–inducible T7 promoter and neomycin / kanamycin resistance ORFs. Therefore, positive clones harboring this plasmid were selected with kanamycin and IPTG. Pullulanase and protease recombinant clones derived from

Shewanella arctica sp. nov. clones bank were screened on LB / kanamycin (50 µg /ml) / IPTG (1 mM) agar medium. 0.5 % (w/v) red - pullulan (Megazyme, Ireland) and 5 % (w/v) AZO–casein (Sigma) were added to select active recombinant clones of pullulanase and protease, respectively. Incubation of screening plates was carried out at 30 °C for 3–4 days. The activity was observed when clearing zones (diffusion halos) around the active clones were seen as a result of red - pullulan or AZO–casein degradation.

2.2.4 Plasmid isolation, sequencing and phylogenetic analysis

Plasmids were isolated from the selected active clones using NucleoSpin plasmid kit (Machereg–Nagel GmbH & Co. KG) and the insert size was determined by restriction analysis technique. DNA sequence was analyzed on an ABI automatic sequencer according to the Sanger method. Primer walking technique was used to determine the sequence of the inserts containing the genes of interest. ORFs and genes were determined by analyzing the inserts sequences with Artemis v5 software and ORF finder online by NCBI database. The BLAST and FASTA algorithms were used to search the databases (Altschul *et al.*, 1990). Signal sequence prediction was carried out using SignalP 3.0 online software.

BLAST analysis was performed by NCBI online database to determine the phylogenetic group to which the new genes were most closely related. Reference amino acid sequences utilized in phylogenetic analysis were retrieved from NCBI database and aligned with the selected genes using CLUSTAL W (1.83) software. The phylogenetic analyses were performed with neighbour joining method by using software from PHYLIP, version 3.57c (Felsenstien, 1995).The phylogenetic tree was drawn using Tree View 32 software.

2.2.5 Protease ORFs/genes subcloning and expression

PCR amplification was carried out using the Expand HiFi PCR kit (Roche Diagnostics, Germany) and temperature gradient thermoblock PCR system (Biometra, Germany) with the following temperature profile: 94 °C for 5 min and 30 cycles of 94 °C for 1 min, 59 ± 5 °C for 1min and 68 °C for 2 min, and then 68 °C for 15 min. The subcloning of PCR–amplified fragments encoding protease genes was carried out by using Acceptor vector kit (Novagen) where, the PCR–fragments were ligated into the expression vector pETBlue-1 and transformed into *E. coli* Tuner (DE3) pLacl expression competent cells.

The plasmid (Fig. 2.1) harbors an IPTG–inducible T7 promoter/terminator and expresses protein with carbenicillin and chloramphenicol. Protease activity could be detected, as clearing zones around the active clones, after 2 -3 days growth of the subclones on LB-carbenicillin (50 μl/ml)-chloramphenicol (34 μg / ml) agar plates containing 5 % (w/v) AZO–casein at 30 °C.

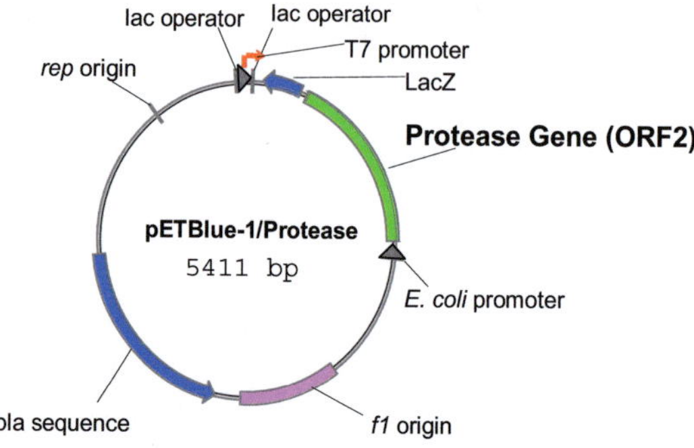

Fig. 2-1: The pETBlue-1/Protease plasmid (5411 bp). Harbor an IPTG-inducible T7 promoter/terminator and the protease gene (ORF2) (1935 bp).

2.3 Purification and characterization of native and recombinant type I pullulanase from *Shewanella arctica.* sp. nov.

2.3.1 Bacterial strain cultivation and native crude enzyme extraction

Cells of *Shewanella arctica* sp. nov. were harvested from 5 l cell culture sample by centrifugation (producing 80 g wet weight cell mass), washed 3 times with phosphate-buffered saline (PBS) (137 mM NaCl, 2.7 mM KCl, 10 mM Na_2HPO_4, 2 mM KH_2PO_4 and final pH of 7.4), and resuspended in buffer (20 mM Tris-HCl – 100 mM NaCl – 1% (v/v) Triton X-100, pH 8) in a constant ratio of 1:5 wet weight cell mass (g) to buffer volume (ml). The resuspended cells were disrupted with a

French press (LM-AMINCO, Spectronic instruments) 3 times at 2500 psi, and the suspension was centrifuged at 10,000g for 30 min at 4 °C. The supernatant produced was used as the native enzyme crude extract.

2.3.2 Enzyme assay and gel electrophoretic analysis

Pullulanase activity was determined by measuring the amount of reducing sugar released during incubation with pullulan (molecular weight 200,000; ICN, Merkenheim, Germany). To 50 µl of 1 % (w/v) pullulan dissolved in 50 mM Tris-HCl buffer (pH 7), 50 or 100 µl of enzyme solution was added, and the samples were incubated at different temperature for 30 to 60 min. The reaction was stopped by cooling on ice, and the amount of reducing sugar released was determined by the dinitrosalicylic acid method (Bernfeld, 1955). Sample blanks were used to correct for the nonenzymatic release of reducing sugar. One unit of pullulanase activity is defined as the amount of enzyme that releases 1 µmol of reducing sugars (with maltose as standard) per min under the assay conditions specified. Protein concentration was determined according to Bradford method (Bradford, 1976).

SDS high- and low-molecular-weight markers (Amersham Biosciences, Germany) were used in order to determine the apparent molecular weight of the samples. SDS-PAGE as described by Laemmli (1970) (Laemmli, 1970) was routinely performed under reducing conditions with 8 % polyacrylamide gels by using a Bio-Rad mini-SUB CELL GT II electrophoresis unit (Bio-Rad, Germany). Samples were mixed with loading buffer, without heat treatment, before loading. After electrophoresis, the gels were rinsed in 0.5 % (v/v) Triton X-100 at 4 °C for 30 min to remove the SDS. Zymogram staining for pullulanase activity was performed by staining the rinsed gel with 5 % (w/v) red-pullulan (in water) (Megayume, Ireland) for 6 hours at 4 °C. The active band appeared (as clearing zone corresponding to red-pullulan hydrolysis) after incubating the stained gel for 30 – 60 min under optimal assay conditions. Gels that were not used for activity staining were soaked in coomassie blue staining solution (0.15 % (w/v) coomassie brilliant blue R-250 in 45.5 % (v/v) methanol-8 % (v/v) acetic acid) for 30 min, this was followed by destaining for 3 hours in 25 % (v/v) ethanol – 35 % (v/v) acetic acid.

2.3.3 Purification of native pullulanase from *Shewanella arctica* sp. nov. and N-terminal sequencing

All purification steps were carried out at room temperature. The native pullulanase crude extracts from 80 g (wet weight) of *Shewanella arctica* sp. nov. cell mass (as described in section 2.4.1) was dialyzed against 100 volumes of 50 mM Tris-HCl, pH 8. The crude extract was then loaded onto a Q-sepharose column (2.5 by 20 cm) equilibrated with 50 mM Tris-HCl, pH 8, washed with the same buffer until no absorbance at 280 nm was detectable. The column was washed with the same buffer, and proteins were eluted with a linear 0–1 M NaCl gradient with the same buffer. Active fractions were pooled and concentrated by ultrafiltration. The enzyme solution was adjusted with ammonium sulfate to a final concentration of 1M and applied at a flow rate of 20 ml/h to a Phenyl-Sepharose column (2.5 by 20 cm) that was equilibrated with 50 mM Tris-HCl, pH 8 containing 1M ammonium sulfate. After the column was washed with 50 ml of the same buffer, the pullulanase was eluted with a linear 1 to 0 M ammonium sulfate gradient in the same buffer. Active fractions were pooled and dialyzed against 20 mM phosphate buffer (pH 7). The enzyme solution was then loaded onto a Hydroxylapatite column (GHT Ceramic) (2.5 by 20 cm) equilibrated with 5 mM phosphate buffer (pH 7). Pullulanase was eluted with a linear 5-200 mM phosphate gradient in the same buffer. Active fractions were pooled and dialyzed against 50 mM Tris-HCl containing 0.15 M NaCl (pH 8). The enzyme solution was then applied to a Superdex S-200 column (2.6 by 96 cm; Amersham Biotech Inc., Germany) equilibrated with 50 mM Tris-HCl-0.15 M NaCl buffer (pH 8). Elution was performed with the same buffer and 2-ml fractions were collected at a flow rate of 2 ml/min. Active fractions were pooled, concentrated by Ultrafiltration (10 kDa cut off) and dialyzed against 25 mM Tris-HCl (pH 8). The enzyme solution was then applied to a Mono Q HR 5/5 column (Pharmacia LKB, Germany), which was equilibrated with 50 mM Tris-HCl (pH 8). Pullulanase was eluted with a linear 0 to 0.5 M NaCl gradient in the same buffer. Fractions containing active pullulanase were analyzed on a sodium dodecyl sulfate (SDS)-8 % polyacrylamide gel electrophoresis (PAGE) and used for the N-terminal sequence determination.

N-terminal protein sequencing was carried out by automated Edman degradation with electroblotted samples on PDVF membrane analyzed on Pulsed liquid protein (model 492, PE-Biosystem, USA) connected online to an HPLC apparatus (HP 1100 Hewleh Packard) for phenylthiohylantion derivative identification.

2.3.4 Optimization of the expression of the recombinant pullulanase from *E. coli* XLOLR

Optimization of expression conditions of the recombinant pullulanase from *E. coli* XLOLR was carried out by growing the cells on LB-kanamycin (50 µg/ml) liquid medium, in the presence and absence of IPTG (1 mM final concentration), at two incubation temperatures (30 and 37 °C), and incubation times (between 6 to 52 hours). The enzyme crude solution was extracted from the recombinant *E. coli* XLOLR and the pullulolytic activity was tested as described in the enzyme assay (2.4.2). The extraction of the crude enzyme solution was carried out by harvesting the cells by centrifugation. The cells were resuspended in buffer (20 mM Tris-HCl / 100 mM NaCl / 1 % (v/v) triton X-100, pH 7) by keeping a constant ratio of 1:5 cell wet weight (g) to buffer volume (ml). The biomass was disrupted by French Press (3 times at 2500 psi), and finally cell debris was removed by centrifugation (12.000g, 20 min, 4 °C).

2.3.5 Purification of recombinant pullulanase from *E. coli* XLOLR

All purification steps were carried out at room temperature. Extraction was performed from 40g *E. coli* XLOLR wet weight as described in section 2.4.4. The enzyme crude extract was dialyzed against 100 volumes of 50 mM Tris-HCl (pH 8). The enzyme solution was then loaded onto a Q-Sepharose column (2.5 by 25 cm), equilibrated with 50 mM Tris-HCl buffer (pH 8) until no absorbance at 280 nm was detectable. The column was washed with the same buffer (pH 8), and proteins were eluted with a linear 0 -1 M NaCl gradient in the same buffer. Active fractions were pooled, concentrated by Ultrafiltration and dialyzed against 20 mM phosphate buffer (pH 7). The enzyme solution was then loaded onto a hydroxylapatite (CHT Ceramic) column (2.5 by 20 cm), equilibrated with 20 mM phosphate buffer (pH 7). After the column was washed with the same buffer, proteins were eluted with a linear 5-200 mM phosphate gradient in the same buffer. Active fractions were pooled and dialyzed against 50 mM Tris-HCl (pH 7) containing 0.1 M NaCl. The enzyme solution was then applied to a Superdex S-200 prep column (2.6 by 96 cm; Amersham Biotech Inc., Germany) that was equilibrated with 50 mM Tris-HCl - 0.10 M NaCl buffer (pH 7). Elution was performed with the same buffer and 2-ml fractions were collected at a flow rate of 2 ml/min. Active fractions were pooled, concentrated by Ultrafiltration (cut off, 10 kDa) and dialyzed against 20 mM Tris-HCl (pH 7). The enzyme solution was then applied to a Mono Q HR 5/5 column (Pharmacia LKB, Germany), which was equilibrated with 50 mm Tris-HCl (pH 7). Pullulanase was eluted with a linear 0-0.5 M NaCl gradient in the same buffer. Fractions containing active pullulanase were analyzed on a sodium dodecyl sulfate (SDS) (8 %) polyacrylamide gel and activity gels were performed as described in section 2.4.2.

2.3.6 Influence of temperature and pH on native and recombinant pullulanase activity

In order to study the influence of temperature and pH, experiments were carried out with the purified native enzyme (3 U/mg) and with the hydroxylapatite pooled active enzyme fractions (2.77 U/mg). To determine the influence of temperature on the enzymatic activity, samples were incubated at temperatures from 4 to 55 °C for 60 min (pH 7). The influence of pH on the enzymatic activity was determined at 45 °C for 60 min by using the assay protocol described in section 2.4.2. 50 mM Tris-HCl buffer was used to obtain pH range from 6 to 13 and sodium acetate buffer to obtain pH range from 2 to 5.

2.3.7 Thermostability of the recombinant pullulanase

Thermostability studies were carried out with the active fractions recombinant enzyme (2.77 U/mg). Thermostability was investigated after incubation of the enzyme samples at low temperatures (-80, -20 and 4 °C) for 0 to 22 days and high temperatures (40, 50 and 65 °C) for 0 to 6 hours (pH 7). After various incubation time intervals, samples were withdrawn and clarified by centrifugation, and the enzymatic activity was measured at 45 °C, pH 7 and for 60 min as described in section 2.4.2.

2.3.8 Effects of metal ions and other reagents on native and recombinant pullulanase activity

The effect of metal ions and several reagents on the pullulanase activity was investigated on the purified native enzyme (3 U/mg) and from recombinant enzyme (2.77 U/mg). The effects of various substances on native and recombinant pullulanase activity were examined after incubation of native and recombinant enzymes with metal ions and other reagents at various concentrations at room temperature for 60 min. Samples were withdrawn and tested for pullulanase activity at 45 °C, pH 7 and for 60 min as described in section 2.4.2. The metal ions tested were (1 mM); $MnCl_2$, Fe_3Cl_2, $CoCl$, $CuCl_2$, $NiCl_2$, $MgCl_2$, $ZnCl_2$ and EDTA (ethylenediamine). The reagents were (all in final concentration); SDS (sodiumdodecysulfate) 1 % (w/v), urea 2 and 6 M, guanidine 0.4 and 0.8 M, α, β, and γ- cylodextrin 0.025 % (w/v), N-bromosuccinide 0.01 % (w/v), DTT (dithiothreitol) 10 mM and idoacetamide 10 and 20 mM.

2.3.9 Substrate specificity and kinetic studies

Substrate specificity was studied on the purified native enzyme (3 U/mg) and recombinant enzyme (2.77 U/mg). In order to study the substrate specificity, the following substrates in additional to pullulane were used: amylopectin, starch, or amylose. Kinetic studies were performed with the recombinant enzyme (2.77 U/mg). In order to calculate the K_m and V_{max} the enzyme activity was tested as described in the enzyme assay protocol (2.4.2) but with various concentrations of pullulan (0.1 to 0.7 % (w/v)). The activity was tested at 45 °C, pH 7 and for 60 min

2.3.10   Characterization of the hydrolysis products of the recombinant pullulanase

The hydrolysis products arising from the action of pullulanase on linear and branched polysaccharides were analyzed by high performance liquid chromatography (HPLC) with an Aminex HPX-42 a column (300 by 78 mm) (Bio-Rad, Hercules, Calif). Double-distilled water was used as the mobile phase at a flow rate of 0.3 ml/min (Rudiger *et al.*, 1995). The recombinant enzyme (2.77 U/mg) was incubated with 0.5 % (w/v) pullulan at 45 °C for 60 min. In order to distinguish maltotriose (only α-1, 4 bonds) from panose or isopanose (α-1,4 and α-1,6 bonds) the incubation was performed with α-glucosidase from yeast (10 U/mg final concentration) in a 50 mM potassium phosphate buffer (pH 7). Incubation of α-glucosidase with the recombinant enzyme was performed by incubating the recombinant pullulanase (2.77 U/mg) with 0.5 % (w/v) pullulan at 45 °C for 60 min and another 60 min with α-glucosidase at 37 °C. DP3 (maltotriose) was also incubated as a control with α-glucosidase at 37 °C for 60 min. α-glucosidase is capable of hydrolyzing α-1,4 but not α-1,6 linkages in short-chain oligosaccharides. Samples were withdrawn and the reaction was stopped by incubation of the mixtures on ice. DP1 (glucose), DP2 (maltose), DP3 (maltotriose) and DP4 were used as standards.

2.4 Purification and characterization of recombinant subtilisin-like serine protease, from *Shewanella arctica.* sp. nov.

2.4.1 Enzyme assay and gel electrophoretic analysis

Proteolytic activity was measured by a modified method of Kunitz (1947) (Kunitz, 1947). Recombinant enzyme fractions (10 µl) were added to 490 µl of 0.25 % (w/v) casein (Hammersten, Merck, Germany) in 120 mM universal buffer, pH 8 (Britton & Robinson, 1931), to give a final volume of 500 µl. The samples were incubated at 60 °C for 10 min. The reaction was stopped by adding 500 µl of 10 % (w/v) trichloracetic acid and the mixture was centrifuged at 13,000 rpm (Heraeus sepatech, Germany) for 10 min. The absorbance of the supernatant was determined at 280 nm. Sample blanks were used to correct the nonenzymatic release of aromatic amino acids. One unit of protease was defined as the amount of enzyme that releases 1 µ mol aromatic amino acids (with tyrosine as standard) per minute under the assay conditions specified. Protein concentration was determined according to Bradford method (1976) (Bradford, 1976).

SDS low–molecular–weight marker (Amersham Biosciences, Germany) was used in order to determine the apparent molecular weight of the samples. SDS-PAGE as described by Laemmli (1970) (Laemmli, 1970) was routinely performed under reducing conditions with 8 % polyacrylamide gels by using Bio-Rad mini-Sub CELL Gt II electrophoresis unit (Bio-Rad, Germany). Samples were mixed with loading buffer without heat treatment before loading. After electrophoresis the gels were rinsed in 0.5 % (v/v) triton X-100 at 4 °C for 30 min to remove the SDS. Zymogram staining for protease activity was performed by staining the rinsed gel with 5 % (w/v) AZO-casein (in water) (Sigma) for 60 min at 4 °C. The active bands appeared (as clearing zones, corresponding to AZO-casein hydrolysis) after incubating the stained gel for 5-10 min under optimal assay conditions and staining it with coomassie brilliant blue for 15 min at room temperature. Gels that were not used for activity staining were soaked in coomassie blue staining solution (0.15 % (w/v) comassie brilliant blue R-250 in 45.5 % (v/v) methanol-8 % (v/v) acetic acid) for 30 min. This was followed by destaining for 3 hours in 25 % (v/v) ethanol–35 % (v/v) acetic acid.

2.4.2 Expression optimization and preparation of crude extract, of the subcloned protease from *E. coli* Tuner (DE3) pLacl

Optimization of expression conditions for the recombinant subcloned protease from *E. coli* Tuner (DE3) pLacl was carried out on LB-carbenicillin (50 µg/ml)-chloramphenicol (34 µg/ml) liquid medium. Protein expression was induced by different IPTG concentrations (0-5 mM final concentration) at various cells optical densities (OD_{600nm} = 0.6 to 1.2). Cells were grown at 30 and 37 °C and up to 60 hours after induction. Crude enzyme solution was extracted from the *E. coli* cells and protealytic activity was tested as described in the enzyme assay protocol section 2.5.1. The extraction of the crude enzyme solution was carried out by cells harvesting by centrifugation, resuspending in buffer

(20 mM Tris-HCl / 100 mM NaCl / 1 % (v/v) Triton X-100, pH 8) with a constant ratio of 1 to 5 cell wet weight (g) to buffer volume (ml), cells disrupting by French Press (3 times at 2500 psi), and cell debris removing by centrifugation (12.000g, 20 min, 4 °C).

2.4.3 Purification of recombinant protease from *E. coli* Tuner (DE3) pLacl

All purification steps were carried out at room temperature. The purification was carried out from 40 g *E. coli* Tuner (DE3)pLacI wet weight (extraction was performed as described in section 2.5.2). The enzyme crude solution was dialyzed against 100 volumes of 50 mM Tris-HCl (pH 8). The enzyme solution was then loaded onto a Q-Sepharose column (2.5 by 25 cm), equilibrated with 50 mM Tris-HCl buffer (pH 8) until no absorbance at 280 nm was detectable. The column was washed with the same buffer (pH 8), and proteins were eluted with a linear 0-1 M NaCl gradient with the same buffer. Active protease fractions were pooled, dialyzed against 50 mM Tris-HCl (pH 8), analyzed on a sodium dodecyl sulfate (SDS)-8 % polyacrylamide gel electrophoresis (PAGE), and activity gels were performed as described in section 2.5.1. This purified enzyme (0.52 U/mg) was used for further studies.

2.4.4 Influence of temperature and pH on recombinant protease activity and thermostability

In order to study the influence of temperature, pH and thermostability, experiments were carried out with the purified recombinant enzyme (0.52 U/mg). To determine the influence of temperature on enzyme activity, samples were incubated at temperatures from 0 to 100 °C for 10 min and pH 8. The influence of pH on the enzymatic activity was determined by using the assay protocol described in section 2.5.1 in 120 mM universal buffer but in various values of pH (6 – 10), and at 60 °C for 10 min. Thermostability was investigated after the incubation of enzyme samples at low temperatures (-80, -20, 4 °C, and room temperature) for 35 days, and at high temperatures (40, 60, 70, 80 and 90 °C) for 45 hours (pH 8). After various time intervals, samples were withdrawn and clarified by centrifugation and the enzymatic activity was measured at 60 °C, pH 8 and for 10 min as described in section 2.5.1.

2.4.5 Effects of metal ions, reagents and inhibitors on protease activity

The effects of metal ions, other reagents and several protease inhibitors on protease activity were investigated by using the purified recombinant enzyme (0.52 U/mg). The effects of various substances on enzymatic activity were examined after incubating the enzyme with metal ions, other reagents and protease inhibitors at various concentrations at room temperature for 60 min. Samples were withdrawn and tested for protease activity at 60 °C, pH 8 and for 10 min as described in section 2.5.1. The activity without metal ions, reagents or inhibitors was considered as 100

%. The metal ions tested were (1 and 5 mM final concentration): $ZnSO_4$, Fe_3Cl_3 x6H$_2$O, $CoCl_2$ x6H$_2$O, Na_2SrO_3 x5H$_2$O, $NaWo^4$, $NiCl_2$ x6H$_2$O, $MgCl_2$, $MnCl_3$ xH$_2$O, $CuSO_4$ x5H$_2$O, $CrCl_3$ x6H$_2$O, $AgNO_3$, and $CaCl_2$ x2H$_2$O. The reagents tested were: CHAPS 1 and 5 mM, SDS (sodium dodecysulfate) 0.5 % (w/v), Triton X-100 0.5 % (v/v), isopropanol 2 % (v/v), DMF (dimethylformamide) 2 % (v/v), DMSO (dimethyl sulfoxide) 5 % (v/v), acetonitril 10 and 20 % (w/v), urea 2 and 6 M , DTT (dithiothreitol) 0.1 % (w/v), and β-mercaptoethanol 0.1 and 0.5 % (v/v). The protease inhibitors tested (Roche Diagnostics, Germany): antipain-dihydrochloride 1 and 5 mM, bestatin 1 and 5 mM, chymostatin 1 and 5 mM, E-64 1 and 5 mM, leupeptin 1 and 5 mM, pepstatin 2.5 mM, phosphoramidon 1 and 5 mM, pefabloc SC (4-(2-aminoethyl)-benzylsulfluoride) 1 and 5 mM, EDTA (ethylenediamine tetraacetic acid) 1 and 5 mM, aspotinin 1 mM, PMSF (phenylmethysulfonyl flouride) 10 and 20 mM.

2.4.6 Kinetic studies

Kinetic studies were performed with the purified recombinant protease (0.52 U/mg). In order to calculate the K_m and V_{max}, the enzyme activity was tested as described in the enzyme assay protocol (2.5.1) but with various concentrations of casein (0.1 to 0.7 % (w/v)). The activity was tested at 60 °C, pH 8 and for 10 min.

3 RESULTS

3.1 *Ilyobacter psychrophilus* sp. nov., a novel anaerobic, psychrophilic bacterium isolated from Spitsbergen

3.1.1 Physiological and morphological characteristics

Enrichment cultures (pH 7) containing yeast extract, peptone and glucose inoculated with seawater sample from Spitsbergen showed bacterial growth after one week of anaerobic incubation at 4 °C (as described in section 2.1.1) Microscopy revealed the presence of short rod cells. After a number of serial dilutions, only one culture was shown to exhibit the same characteristics and was selected as the culture for the strain (FQ50).

Cells of the strain FQ50 were found to be gram negative. They were short rods (0.2-0.3 µm wide and 0.5-0.7 µm long) and non motile. They occurred singly, in pairs or as short chains. Spores could never be detected. A comparison of the morphological characteristics of the strain FQ50 and the related strains (belonging to the *Ilyobacter-Propionigenium* group): *Ilyobacter tartaricus* (AJ307982.1), *Ilyobacter insuetus* (AJ307980.1), *Ilyobacter polytropus* (AJ307981),

Propionigenium maris (Y16800.1) and *Propionigenium modestum* (X54275.1) is shown in Table 3-1.

Table 3-1: Comparative characteristics of the strain FQ50 and the strains of *Ilyobacter-Propionigenium* group.

1. FQ50, 2. *Ilyobacter tartaricus* [Schink, (1984)], 3. *Ilyobacter polytropus* [Stieb & Schink, (1984)], 4. *Ilyobacter insuetus* [Brune *et al*, (2002)], 5. *Propionigenium maris* [Watson *et al*, (2000) and Janssen & Liesack, (1995)], 6. *Propionigenium modestum* [Schink & Pfennig, (1982)]. +, positive; -, negative; ND, not determined. OD is the optical density at 600 nm.

Character	1	2	3	4	5	6
Cell shape	Short rod	Short rod	Short rod	Cocciod	Short rod	short rod
Cell size (µm):						
Length	0.5 - 0.7	1.2 - 2.5	1.5	1 - 1.5	1.2 - 2.5	0.5 - 2
Diameter	0.2 - 0.3	1 - 1.2	0.7	0.8 -1		0.5 - 0.6
Gram stain	-	-	-	-	-	-
Mobility	-	-	+	-	-	-
Spore formation	-	-	-	-	-	-
Optimum growth temperature (°C)	10	28 - 34	28 - 34	30	34 - 37	33
Optimum pH for growth	7	6.5 - 7.2	7	6- 8	6.9 - 7.7	7.1 - 7.7
Optimum NaCl conscetration for growth (%, w/v)	2	1	1	0.7	0.5	2
Oxygen relationship	anaerobe	anaerobe	anaerobe	anaerobe	anaerobe	anaerobe
DNA G + C content (mol %)	28.7	33.1	32.2	35.7	40	33.9
Reduction of electron acceptors:						
Sulfate (10 mM)	-	-	-	-	-	-
Sulfur (0.1 %)	-	-	-	-	-	-
Thiosulfate (10 mM)	-	-	-	-	-	-
Nitrate (10 mM)	-	-	-	-	-	-
Fermentation product of different substrates:						
Butyrate	+	-	+	-	+	-
Acetate	+	+	+	+	+	+
Lactate	-	-	-	-	+	-
Formate	-	+	+	-	+	-
Ethanol	-	+	+	-	+	-
Propionate	+	-	+	-	+	+
Utilization of:						
Succcinate	-	-	-	-	+	+
Pyruvate	OD 0.4	+	+	-	+	+
Starch	OD 0.6	ND	ND	ND	ND	ND
Glucose	OD 0.75	+	+	-	+	-
Fumarate	-	-	+	-	+	+
Citrate	OD 0.17	+	+	-	+	-
Fructose	OD 0.59	+	+	-	+	-
Maltose	OD 0.22	-	-	ND	+	-
Malate	-	-	+	-	-	+
Arabinose	OD 0.14	-	-	ND	-	-
Crotonate	-	ND	+	-	-	-
Oxaloacetate	OD 0.22	+	+	ND	-	+
L-aspartate	-	ND	-	-	-	+
Lysine	-	ND	-	ND	+	-
Threonine	-	ND	-	-	+	-
Glutamate	OD 0.18	ND	-	ND	+	-
Cysteine	OD 0.12	ND	-	ND	+	-
Yest extact	OD 0.8	-	-	ND	+	-
L-tartate	-	+	-	-	-	-
D-tartate	-	-	-	-	-	-
Glycerol	OD 0.23	-	+	-	-	-
Xylase	OD 0.13	-	-	-	-	-
3 hydroxybutyrate	OD 0.11	-	+	-	+	-

No growth was observed under aerobic condition with yeast extract (2.5 g^{-1}), peptone (2.5 g^{-1}) and glucose ($1g^{-1}$). Growth occurred only under anaerobic conditions between 4 and 20 °C, the optimal growth was optimum at 10 °C where the growth rate reached its maximum (Fig. 3-1A). No growth was observed at 25 °C or above. The strain FQ50 required seawater and grew well at salt (NaCl) concentration of 1–6 % (w/v) with optimum at 2 % NaCl (w/v) (Fig. 3-1B). The pH range for growth was 5–9, with an optimum at pH 7 (Fig. 3-1C). Under the optimum conditions using yeast extract (2.5 g^{-1}), peptone (2.5 g^{-1}) and glucose ($1g^{-1}$) the growth rate was 0.397 h^{-1}. The growth conditions of the strain FQ50 and the related strains are shown in Table 3-1.

3.1.2 Substrate utilization

The strain FQ50 grew on a variety of substrates. Growth was observed on yeast extract, pyruvate, starch, glucose, citrate, fractose, maltose, arabinose, oxaloacetate, glutamate, cysteine, glycerol, xylose or 3-hydroxybutyrate (Table 1-3). No growth was observed on succinate, fumarate, malate, crotonate, L–aspartate, lysine, theonine, L–tartate or D-tartate. The strain FQ50 fermented glucose, starch and pyruvate resulting in the formation of butyrate, acetate and propionate. Lactate, formate and ethanol could not be detected as fermentation products. Sulfate, sulfur, thiosulfate and nitrate were tested as possible electron acceptors during the growth of the stain FQ50 on glucose. No effect on growth or on the formation of fermentation products was observed. At the end of growth neither the concentration of electron acceptors was decreased nor reduction products were detected. Substrates utilization, fermentation products and electron acceptor reduction of the new strain FQ50 and the related strains are listed in Table 3-1.

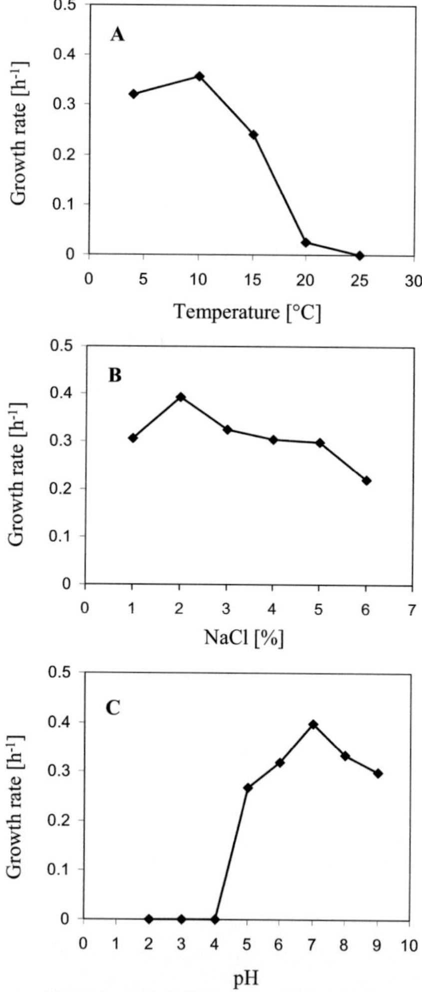

Fig. 3-1: (A) Effect of the temperature at pH 7 and NaCl concentration 2 % (w/v) on the growth of the strain FQ50. (B) Effect of the NaCl concentration at 10 °C and pH 7 on the growth of the strain FQ50. (C) Effect of the pH at 10 °C and NaCl concentration 2 % (w/v) on the growth of the strain FQ50.

3.1.3 Fatty acid analysis

The fatty acid methyl ester (FAME) composition of the strain FQ50 is shown in Table 3-2. The FAME profile displays only those fatty acids comprising $\geq$ 0.1 % of the total. Straight chain saturated FAMEs were 28.17 % total, terminally branched saturated FAMEs were 3.21 % total and monounsaturated FAMEs were 22.13 % total. 16:0 straight chain saturated FAME (19.7 %) and 15:1 ω6c monounsaturated FAME (13.58 %) were the most abundant FAMEs found. The FAME composition of the related strains were not reported (Brune *et al.*, 2002; Janssen & Liesack, 1995; Schink, 1984; Schink & Pfenning, 1982; Stieb & Schink, 1984; Watson *et al.*, 2000).

Table 3-2: Fatty acid composition (%) of the isolated strain FQ50.

Fatty acids	[%]
Sraight-chain fatty acids	
12:00	0.74
14:00	2.33
15:00	2.88
16:00	19.71
17:00	1.81
18:00	0.7
sum	**28.17**
Terminally branched saturated fatty acids	
11:0-iso	0.16
12:0-iso	1.64
15:0-iso	0.69
16:0-iso	0.72
sum	**3.21**
Monounsaturated fatty acids	
14:1 ω5c	3.24
15:1 ω6c	13.58
15:1 ω8c	0.76
16:1 ω5c	0.75
17:1 ω6c	0.8
17:1 ω8c	2.21
18:1 ω7c	0.79
sum	**22.13**

3.1.4 DNA base composition and phylogenetic position

The G + C content of the DNA from the strain FQ50 was 28.7 mol %. The 16S rDNA sequence of the strain FQ50 (1.409 kb) was analyzed and compared with all currently available sequences of organisms belonging to the *Fusobacteriaceae* family (Fig. 3-2). The closest relationship was with, *Ilyobacter polytropus* and *Propionigenium maris,* members of *Ilyobacter-Propionigenium* group with an identity on 16S rDNA level of 92 %.

3.1.5 Description of *Ilyobacter psychrophilus* sp. nov.

Ilyobacter psychrophilus (Psychrophilus adj. derived from Latin meaning cold loving) cells are short rod shaped, gram negative, 0.5–0.7 μm long and 0.2–0.3 μm wide, occurred singly, in pairs or as short chains, non motile, non spore forming and strictly anaerobic. Temperature range for growth is 0–20°C, with an optimum at 10 °C. Range of pH for growth 5-9, with an optimum at pH 7. Growth occurs in 1–6 % NaCl (w/v) concentration, with an optimum at 2 % (w/v). The strain grows with pyruvate, starch, yeast extract, peptone, glucose, citrate, fructose, maltose, arabinose, oxaloacetate, glutamate, cysteine, glycerol, xylose or 3 hydroxybutyrate. Growth is observed in the presence of sulfate, sulfur, thiosulfate or nitrate. Fermentation products on glucose, starch or pyruvate are butyrate, acetate and propionate. The fatty acid methyl esters (FAMEs) are composed of 28.17 % straight chain saturated FAME, 3.21 % terminally branched saturated FAME and 22.13 % monounsaturated FAME. Phylogenetic analysis reveals as close relationship to *Ilyobacter polytropus* (92 %). The 16S rDNA is 1.409 kb and DNA base ratio is 28.7 mol % G + C. Habitat: artic seawater. The strain FQ50, was isolated from seawater sample taken from Spitsbergen.

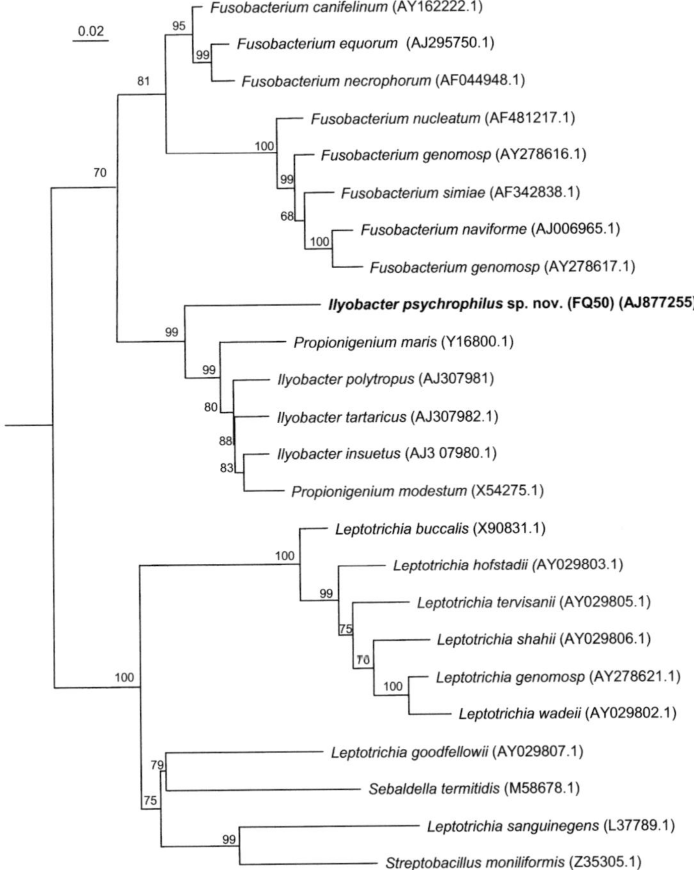

Fig. 3-2: Pylogenetic dendrogram based on 16S rDNA gene sequence comparison indicating the position of the new strain *Ilyobacter psychrophilus* sp. nov. FQ50 within the family *Fusobacteriaceae.* preformed by the neighbour-joining method using software from PHYLIP, version 3.57c; the DNADIST program with Kimura-2 factor was used to compute the pairwise evolutionary distances for the above aligned sequences, the topology of the phylogenetic tree was evaluated by performing a bootstrap (algorithm version 3.6 b) with 1000 bootstrapped trials. The tree was draw using Tree View 32 software. Bar corresponds to 2 nucleotide substitutions per 100 nucleotides.

3.2 *Shewanella arctica* sp. nov., a novel psychrophilic bacterium isolated from Spitsbergen

3.2.1 Physiological and morphological characteristics

Enrichment cultures (pH 7) containing glucose inoculated with seawater sample from Spitsbergen showed bacterial growth after one week of incubation at 4 °C (as described in section 2.1.1). Microscopy revealed the presence of straight rod cells. After a number of transfer on marine medium agar plates at 15 °C, one culture was shown to exhibit the same characteristics and was selected as the culture for the strain (40-3).

Cells of the strain 40-3 were found to be gram negative. They were straight or curved rod - shaped (0.4-0.6 μm wide and 2-3 μm long). They occurred singly. Spores could never be detected. The morphological characteristics of the strain 40-3 and the related strains (belonging to the genus *Shewanella*) are listed in Table 3-3.

No growth was observed under anaerobic condition with glucose. Growth occurred only under aerobic conditions between 4 and 25°C. Optimal growth was optioned at 10-15°C where the growth rate reached its maximum (Fig. 3-3A). No growth was observed above 25°C. The strain 40-3 required seawater and grew well at salt (NaCl) concentration of 0–10 % (w/v) with optimum at 8-9 % NaCl (w/v) (Fig. 3-3B). The pH range for growth was 6–9, with an optimum at pH 7–8 (Fig. 3-3C). Under the optimum conditions using 2 g^{-1} glucose, the growth rate was

0.48 h^{-1}. The growth conditions of the strain 40-3 and the related strains (belonging to the genus *Shewanella*) are listed in Table 3-3.

Table 3-3: Comparative characteristics of the isolated strain 40 - 3 with related *Shewanella* sp strains

1. Strain 40-3, **2.** *Shewanella putrefaciens* [Venkateswaran *et al.*, (1999)]; **3.** *Shewanella baltica* [Ziemke *et al.*, (1998)]; **4.** *Shewanella frigidimarina* [Bowman *et al.*, (1997)]; **5.** *Shewanella pacifica* [Ivanova *et al.*, (2004)]; **6.** *Shewanella gaetbuli* [Yoon *et al.*, (2004)]; **7.** *Shewanella waksmanii* [Ivanova *et al.*, (2003)]. +, positive; -, negative; ND, not determined.

Character	1	2	3	4	5	6	7
Cell shape	Straight rod	Straight rod	Straight rod	Straight rod	Straight rod	Straight rod	Straight rod
Gram stain	-	-	-	-	-	-	-
Spore formation	-	-	-	-	-	-	-
Optimum growth temperature (°C)	10 -15	25 - 33	Growth at 4	20 - 22	20 - 25	30	20 - 22
Optimum pH for growth	7 - 8	7 - 8	ND	ND	ND	7 - 8	7.5
Optimum NaCl conscetration for growth (%, w/v)	8 - 9	0 - 6	ND	0 - 6	0.5 - 6	3 - 6	1 - 6
DNA G + C content (mol %)	48	47	46	40 - 43	40	42	43
Reduction of:							
NO3- to NO2-	+	+	+	+	+	-	ND
NO2- to N	-	-	ND	ND	ND	-	ND
Production of:							
Amylase	+	-	ND	-	+	+	-
Protease	+	ND	ND	ND	+	ND	ND
Pullulanase	+	ND	ND	ND	ND	ND	ND
Lipase	-	-	+	+	+	+	+
H2S from sodium thiosulfate	+	+	+	+	ND	-	+
Indole (tryptophane)	-	-	-	ND	-	ND	ND
Acetion	-	-	-	ND	-	ND	ND
Utilization of:							
α cyclo-dextrin	+	ND	ND	ND	ND	ND	ND
Dextrin	+	ND	ND	ND	+	ND	ND
Tween 80	+	ND	ND	ND	ND	ND	ND
N-acetyl-D-glucosamine	+	ND	ND	ND	+	ND	-
α-D-glucose	+	ND	ND	ND	ND	ND	+
Maltose	+	+	+	-	+	-	ND
Sucrose	+	+	+	-	ND	-	-
Methyl pyruvate	+	ND	ND	ND	ND	ND	ND
D,L-lactic acid	+	ND	ND	ND	ND	ND	ND
Succiniate	+	+	ND	+	+	-	-
Bromo succiniate	+	ND	ND	ND	ND	ND	ND
Inosine	+	ND	ND	ND	ND	ND	ND
Esculin ferric citrate	+	ND	ND	ND	ND	ND	ND
L-arabinose	+	ND	ND	ND	+	ND	ND
Potassium gluconate	+	ND	ND	ND	ND	ND	ND
Malic acid	+	ND	+	ND	ND	ND	ND
Trisodium citrate	+	ND	ND	ND	ND	ND	ND
D-galactose	-	+	ND	-	+	-	-
D-fructose	-	-	ND	-	-	-	-
Fumarate	ND	+	ND	+	-	-	-
Lactose	-	+	ND	ND	-	-	-
Citrate	-	-	+	ND	-	-	ND

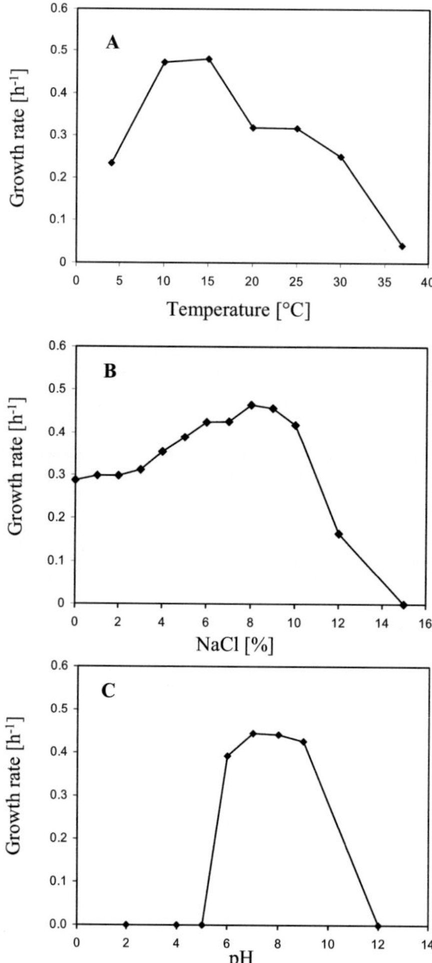

Fig. 3-3: (A) Effect of the temperature at pH 7 and NaCl concentration 2 .8 % (w/v) on the growth of the strain 40-3. (B) Effect of the NaCl concentration at 15 °C and pH 7 on the growth of the strain 40-3. (C) Effect of the pH at 15 °C and NaCl concentration 2.8 % (w/v) on the growth of the strain 40-3.

3.2.2 Substrate spectrum

The strain 40-3 grew on a variety of substrates. Good growth was observed on: α-cyclodextrin, dextrin, tween 80, N-acetyl-D-glucosamine, α-D-glucose, maltose, sucrose, methylpyruvate, D,L-lactic acid, succinic acid, bromo succinic acid, inosine, esculin ferric citrate, L-arabinose, potassium gluconate, malic acid and trisodium citrate. No growth was observed on glycogen, tween 40, N-acetyl-D-galacto-samine, adonitol, capric acid, D-arabitol, cellobiose, I-erythol, D-fructose, L-fructose, D-galactose, gentiobiose, m-inositol, α-D-lactose, lactulose, D-mannitol, D-mannose, D-mellobiose, β-methyl D-glucose, D-psicose, D-raffinose, L-rhamnose, D-sorbitol, D-trehalose, turanose, xylitol, mono-methyl succinate, acetic acid, cis-aconitic acid, citric acid, formic acid, D-galactonic acid lactone, D-galacturonic acid, D-gluconic acid, D-glucosaminic acid, D-glucuronic acid, α-hydroxy-butyric acid, β-hydroxy-butyric acid, γ-hydroxy-butyric acid, β-hydroxy phenylacetic acid, itaconic acid, α-keto butyric acid, α-keto glutaric acid, α-keto valeric acid, malonic acid, propionic acid, quinic acid, D-saccharic acid, seabacic acid, succinamic acid, glucuronamide, alaninamide, D-alanine, L-alanine, L-alanyl-glycine, L-asparagine, L-aspartic acid, L-glutamic acid, glycyl-L-aspartic acid, glycyl-L-glutamic acid, L-histidine, hydroxy L-proline, L-leucine, L-ornithine, L-phenyl-alanine, L-proline, L-pyroglutamic acid, D-serine, L-serine, L-threonine, D,L-carnitine, γ-amino butyric acid, urocanic acid, uridine, thymidine, phenylthylamine, putrescine, 2-aminoethanol, 2,3-butanediol, glycerol, D,L α-glycerol phosphate, glucose 1-phosphate, glucose 6-phosphate or phenylacetic acid. The strain 40–3 was able to produce H_2S. Indole (Tryptophane) and aceton were not produced. The strain was also able to reduce nitrates to nitrites but not nitrites to nitrogen. The ability of the strain 40–3 and related strains, (belonging to the genus *Shewanella*) to grow on various substrates is listed in Table 3-3.

3.2.3 Enzymes production

The strain 40–3 was able to produce several enzymes: amylase, pullulanase, protease, omithine decarboxylase, alkaline phospatase, esterase (C4), esterase/lipase (C8), leucine arylamidase, valine arylamidase, trypsin, α-chymotrpsin, naphthol-AS-BI-phosphohydrolase and N-acetyl-β–glucosaminidase. However, arabinase, arabinoxylanase, HE-cellulase, glucanase, dextranase, galactanase, galactomannanase, β–glucanase, curdlanase xylanase, xyloglucanase, lipase, α- and β–galactosidase, β–glucoronidase, α and β–glucosidase, α–mannosidase, α–fucosidase, acid phosphatase, cystine arylamidase, arginine dihydrolase, lysine decarboxylase, urease and tryptophane deaminase were not detected.

3.2.4 Fatty acid analysis

The FAME composition of the strain 40–3 is shown in Table 3-4. The FAME profiles display only those fatty acids comprising ≥ 0.05 % of the total. Straight chain saturated FAMEs were 17.89 % total, terminally branched saturated FAMEs were 14.85 % total and monounsaturated FAMEs were 17.73 % total. 16:0 straight chain saturated FAME (11.24 %), 15:0-iso terminally branched saturated FAME (9.4 %) and 18:1 ω7c monounsaturated FAME (9.21 %) were the most abundant FAME s found. The FAME composition of the strain 40–3 and related strains (belonging to the genus *Shewanella*) are listed in Table 3-4.

Table 3-4: Comparative fatty acids composition (%) of the new isolated strain 40-3 and related *Shewanella* species

1. Strain 40-3, **2.** *Shewanella putrefaciens* [Venkateswaran *et al.*, (1999)]; **3.** *Shewanella baltica* [Ziemke *et al.*, (1998)]; **4.** *Shewanella frigidimarina* [Bowman *et al.*, (1997)]; **5.** *Shewanella gaetbuli* [Yoon J. H. *et al.*, (2004)]; **6.** *Shewanella waksmanii* [Ivanova E. P. *et al.*, (2003)]. Both the strain 40 – 3 and *Shewanella frigidimarina* produce eicosapentaenoic acid, 20:5ω3. ND, not determined.

Fatty acids	1	2	3	4	5	6
Straight-chain fatty acids						
12:00	3.64	ND	ND	ND	3.1	2
13:00	0.07	ND	ND	ND	ND	ND
14:00	1.8	2.3	2.2	3.7	ND	1.7
15:00	0.31	3.2	7.8	2.5	3.8	5.3
16:00	11.24	19.1	4.3	11.8	8.4	6.2
17:00	0.31	1.5	0.6	1.2	ND	ND
18:00	0.52	2.1	ND	0.1	ND	0.3
Terminally branched saturated fatty acids						
13:0-iso	4.13	2.5	12.4	6.3	9.4	10
14:0-iso	0.18	0.3	1.6	0.6	2	ND
15:0-iso	9.4	21.1	14.3	9	ND	32.5
17:0-iso	1.03	1.7	0.5	0.3	ND	ND
18:0-iso	0.11	ND	ND	ND	ND	ND
Monounsaturated fatty acids						
15:1 ω6c	0.1	0.2	2.2	1.2	ND	ND
16:1 ω6o	0.28	ND	ND	ND	ND	ND
16:1 ω7c		29.6	24.1	51.1	21.1	9.8
16:1 ω9c	2.09	3.5	1.6	2.2	ND	ND
17:1 ω8c	0.89	6.7	11	3	6.4	ND
17:1 ω6c	0.24	0.9	1.4	ND	ND	ND
18:1 ω9c	4.5	3.8	0.3	1.7	ND	ND
18:1 ω7c	9.21	6.6	0.8	5.3	2.5	2
18:1 ω5c	0.35	ND	ND	ND	ND	ND

3.2.5 DNA base composition and DNA - DNA hybridization

The G + C content of the DNA from the strain 40-3 was 48 mol %. The 16S rDNA sequence of the strain 40-3 (1.447 kb) was analyzed and compared to all currently available 16S rDNA sequences of organisms belonging to the genus *Shewanella* (Fig. 3-4). The closest relationship was to *Shewanella putrefaciens* (U91552.1) with an identity on 16S rDNA level of 99 %. However, *Shewanella massilia* (AJ006084.1) and *Shewanella baltica* (AJ000214.1), showed 98% identity, *Shewanella decolorationis* (AJ609571.1) and *Shewanella frigidimarina* (AJ300833.1) showed 97 % identity, *Shewanella pacifica* (AY366086.1) and *Shewanella gaetbuli* (AY190533.1) showed 96% identity and *Shewanella marisflavi* (AY485224.1) and *Shewanella waksmanii* (AY170366.1) showed 95 % identity, on the 16S rDNA level. The DNA–DNA hybridization of the strain 40–3 to the most related strain *Shewanella putrefaciens* (U91552.1, DZM 6067) showed only 50.1 % DNA–DNA similarity.

3.2.6 Description of *Shewanella arctica* sp. nov.

Shewanella arctica (arc' ti.ca. L. fem. adj. *arctica* from the Arctic, referring to the site were the type strain was isolated). Cells are straight or curved rod- shaped, gram negative, 2-3 µm long and 0.4–0.6 µm wide, occur singly, non spore forming and strictly aerobic. Colonies on agar medium are slightly orange, circular, smooth and convex. Temperature range for growth is 4–25 °C, with an optimum between 10 and 15 °C. No growth at 30 °C was detected. Range of pH for growth 6–

9, with an optimum at pH 7–8. Growth from 0-10 % NaCl (w/v) concentrations, with an optimum between 8 and 9 % (w/v). Growth is observed with α-cyclo-dextrin, dextrin, tween 80, N-acetyl-D-glucosamine, α-D-glucose, maltose, sucrose, methylpyruvate, D,L-lactitate, succiniate , bromo succinic acid, inosine, esculin ferric citrate, L-arabinose, potassium gluconate, malic acid and trisodium citrate. In the presence of glucose, H_2S is produced and nitrates are reduced to nitrites. The strain forms the following enzymes: Amylase, pullulanase, protease, ornithine decarboxylase, alkaline phospatase, esterase (C4), esterase lipase (C8), leucine arylamidase, valine arylamidase, trypsin, α-chymotrpsin, naphthol-AS-BI-phosphohydrolase and N-acetyl-β–glucosaminidase. The fatty acid methyl esters (FAMEs) were composed of 17.89 % straight chain saturated FAME, 14,85 % terminally branched saturated FAME and 17.73 % monounsaturated FAME. Phylogenetic analysis reveals as close relationship to *Shewanella putrefaciens* with 99 % 16S rDNA composition identity and 50 % DNA-DNA similarty. The 16S rDNA is 1.447 kb. The DNA base ratio is 48 mol % G + C. Habitat: artic sea water. The strain 40-3 = DSM 16509, was isolated from seawater samples taken from Spitsbergen.

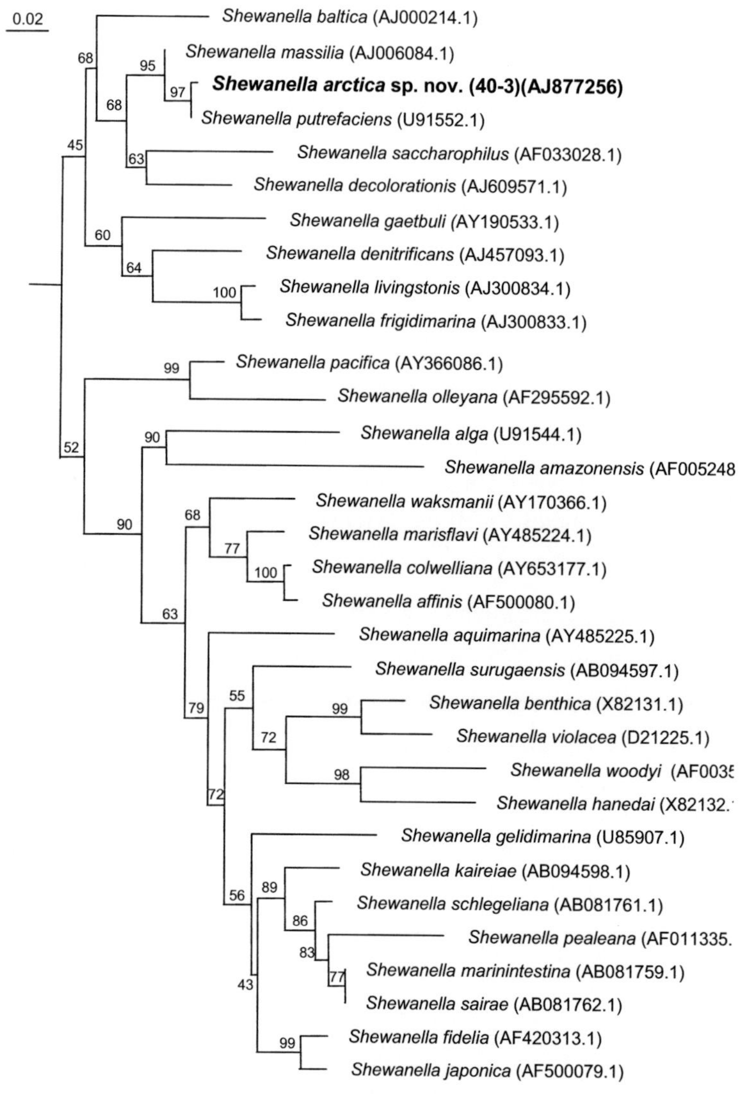

Fig. 3-4: Phylogenetic dendrogram based on 16S rDNA gene sequence comparison indicating the position of the new strain *Shewanella arctica* sp. nov. 40-3 within the genes *Shewanella.* performed by the neighbour-joining method using software from PHYLIP, version 3.57c; the DNADIST program with Kimura-2 factor was used to compute the pairwise evolutionary distances for the above aligned sequences , the topology of the phylogenetic tree was evaluated by performing a bootstrap (algorithm version 3.6 b) with 1000 bootstrapped trials. The tree was draw using Tree View 32 software. Bar correspond to 2 nucleotide substitutions per 100 nucleotides.

3.3 Cloning, sequencing, purification and characterization of a type I pullulanase from *Shewanella arctica.* sp. nov.

3.3.1 Cloning and sequencing of the 6-kb insert encoding pullulanase from *Shewanella arctica* sp. nov.

Hundred-thousend phagemid clones obtained from a λZAP (Stratagen) genome bank were screened for pullulanase activity at 30 °C for 3-4 day as described in materials and methods section 2.3.3. Two clones showed clearing zones on the red-pullulan plates. The insert sizes were 6.091 kb for the first clone and 4 kb for the second one. The 6.091 kb clone was sequenced using primers walking technique. Four forward and reverse walking with 8 constructed primers (Table 3-5) were necessary to determine the complete sequence of the insert. M13 forward and M13 reverse standard primers were used to start the primer walking. Complete sequence revealed a G+C content of 46.82 % and one open reading frame (ORF) as shown in Fig. 3-5. This ORF (4.323 kb) encodes a protein of 1440 amino acids with a predicted molecular mass of 155.9 kDa. The BLAST search by GenBank database showed that this ORF/gene has similarity to known pullulanase.

Table 3-5: Oligonucleotides used for the primer walking* of the 6-kb insert encoding pullulanase gene

walk	Primer direction F = forward R = reverse	Primer position (bp)	Sequences (5$'$ to 3$'$)
1	F	652	AGC CTC CTA CCA AGT AAC G
	R	644	TCG CAT TAT CTG ACA GGT G
2	F	1296	CCG CTC CTT TTG ATA CCA G
	R	1414	TGA CGC ATC AAA CAC AAA G
3	F	1973	GAT GGG GCG AAC ATT ATT G
	R	2148	TTC GCA CCC TCA CTA AAA C
4	F	2597	AAA ACC CAA AAC AAG TCA G
	R	2851	GCG TAA GCA CTC GGT TGT G

* M13 forward and M13 reverse standard primers were used to start the primer walking.

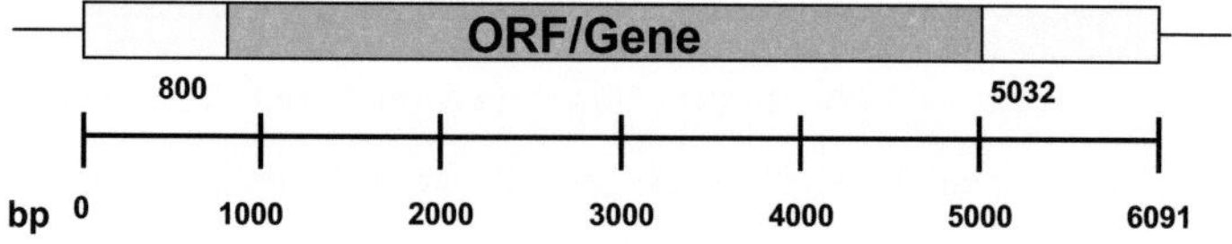

Fig 3-5: Map of the 6-kb insert encoding the pullulanase from *Shewanella arctica*. The whole insert (6.091 kb) revealed a G+C content of 46.82 % and one open reading frame (ORF) encoding the pullulanase gene (marked in gray). The gene, including the signal sequence, 800-bp to 5032-bp. has a size of 4.323 kb, G+C content of 47.97 %, encodes a protein of 1440 amino acids with a predicted molecular mass of 155.9 kDa

3.3.2 Pullulanase gene analysis and phylogenetic position

The pullulanase gene (Fig 3-6) from *Shewanella arctica* sp. nov. has a ATG (M) start codon, a TAG stop codon, is 4.323 kb long, has a G+C content of 47.97 %, and encodes an enzyme with a predicted molecular mass of 155.9 kDa (1440 aa). In addition, a signal sequence consisting of 23 amino acids with predicted cleavage taking place between alanine residue and cysteine residue was found. The hypothetical Shine-Dalgarno-like sequence (AGGAGA) was found at 9 to 14 bp upstream from predicted translational start codon.

The pullulanase gene exhibits 53 % amino acid identity with the type I pullulanase from *Microbulbifer degradans*, 39 % with the type I pullulanase from *Vibrio vulnificus*, 37 % with the type I pullulanase from *Klibsiella pneumoniae* and 34 % with the type I pullulanase from *Chloroflexus aurantiacus*. A comparison of the identity and similarity percentage of the pullulanase amino acids sequence form *Shewanella arctica* sp. nov. with the most related pullulanase 26 different bacterial strains is shown in Table 3-6.

The highly conserved seven-residue motif typical of all type I pullulanase FNWGYDP was detected at position 598 to 604. The four conserved regions present in amylolytic enzymes belonging to glycosyl hydrolase family 13 and typically found in pulluanase, were identified in *Shewanella arctica* sp. nov pullulanase. Comparison of all conserved regions found in the new pullulanase and the most related type I pullulanase from different bacterial strains are presented in Table 3-7 and Table 3-6.

Construction of phylogenetic tree based on amino acids sequences from the *Shewanella arctica* sp. nov. pullulanase and the most related type I pullulanases from the bacterial strains: *Microbulbifer degradans* (ZP_00315635.1), *Vibrio vulnificus* (NP_936115.1), *Klebsiella pneumoniae* (CAA36431.1), *Chloroflexus aurantiacus* (AAN85642.1), *Bdellovibrio bacteriovorus* (NP_968139.1), *Bacillus cereus* (YP_084052.1), *Bacillus thuringiensis* (YP_036824.1), *Anaerobranca horikoshii* (AAP45012.1), *Anaerobranca gottschalkii* (AAS47565.1), *Clostridium perfringens* (BAB81258.1), *Exiguobacterium* sp. (ZP_00182722.2), *Lactobacillus acidophilus* (YP_194553.1), *Geobacillus kaustophilus* (YP_148680.1), *Thermus thermophilus* (BAB62095.1), *Bacillus subtilis* (NP_390871.1), *Bacteroides thetaiotaomicron* (AAC44685.1), *Fervidobacterium pennivorans* (AF096862_1), *Francisella tularensis* (YP_169457.1), *Bacillus halodurans* (BAB06983.1), *Streptococcus pyogenes* (ZP_00365686.1), *Oceanobacillus iheyensis* (NP_691326.1), *Thermotoga maritime* (NP_229641.1), *Streptococcus suis* (ZP_00332302.1), *Streptococcus pneumoniae* (AF217414_1), *Clostridium acetobutylicum* (NP_349286.1), *Mycoplasma mobile* (YP_016097.1) is shown in Fig. 3-7.

gagattaacgatcaagtggtgaacttcgattccgaaaccgtgatg

gactatcaagcctcctaccaagtaacggacggtttaaacgtactg

ttccaagtcaataacctcacagatgaaccgactaagagttacttc

ggcaccgagcagaaaaccggcaccctacagtactttggccgcgaa

tttttcttgggggttcacctatgccctgtaatccattgcatcagtt

agtgtgcttagacgcacagagcctaaatccattacatgctgcgcc

acttaatatcacggcgccttattcgcaatt<u>aggaga</u>aaccactc

```
  1  atgttgagtatctctaatcttattaaactatcggctttagccctg
      M   L   S   I   S   N   L   I   K   L   S   A   L   A   L

 46  agtatgagcttactctcggcctgtggcggcagtgattcatctgaa
      S   M   S   L   L   S   A   C   G   G   S   D   S   S   E

 91  ccgggtaaagtactattaacttgcgatgtgcctatggtgcccgat
      P   G   K   V   L   L   T   C   D   V   P   M   V   P   D

136  gcgacgggcacaagctgcgttgcacctacgcctatcagttgtaag
      A   T   G   T   S   C   V   A   P   T   P   I   S   C   K

181  gcgcctaaattccctgatcctaaaaacgagagttgtataaacggt
      A   P   K   F   P   D   P   K   N   E   S   C   I   N   G

226  ttagatccctcgttacccgcgccaagcgtattcccgcaagccaat
      L   D   P   S   L   P   A   P   S   V   F   P   Q   A   N

271  caagcagtgctttactacaatcgcattaaagataaaaacaatact
      Q   A   V   L   Y   Y   N   R   I   K   D   K   N   N   T

316  gcgagtaacgctgaatatcctaagtacaaattacatacttggaac
      A   S   N   A   E   Y   P   K   Y   K   L   H   T   W   N

361  gatgaaaagtgtgatgcctatgccgctccttttgataccagcgac
      D   E   K   C   D   A   Y   A   A   P   F   D   T   S   D

406  tggagcaacggccatcccttcgatggtattgaccctaattatggc
      W   S   N   G   H   P   F   D   G   I   D   P   N   Y   G

451  gcctattggatccttaatctgaaggaaggttatggtgcctgcggc
      A   Y   W   I   L   N   L   K   E   G   Y   G   A   C   G

496  aactttattgtgcatgccggtacggatgatgcgggtaaagcctta
      N   F   I   V   H   A   G   T   D   D   A   G   K   A   L

541  ggtaaaaacaacttaaccatgtccctggtacaggacgatgcgacg
      G   K   N   N   L   T   M   S   L   V   Q   D   D   A   T

586  tatgttcgcgtgaatttcaccattcacggtgagtcggatgtgtat
      Y   V   R   V   N   F   T   I   H   G   E   S   D   V   Y

631  gagtatccgattctagagacaggtttctctatttcaggtatgagt
      E   Y   P   I   L   E   T   G   F   S   I   S   G   M   S

676  gcccactggttagatcgcaatacctttgtgtggaatcaagatcca
      A   H   W   L   D   R   N   T   F   V   W   N   Q   D   P

721  caggggggtgacaagtgtgaaattgcattattcagctaaggctgat
      Q   G   V   T   S   V   K   L   H   Y   S   A   K   A   D

766  atcgccgccgatgataataatgtgatcagtggcactcaagtgacc
      I   A   A   D   D   N   N   V   I   S   G   T   Q   V   T

811  ttaacgcccgcgaccttaagcgatgcacaaaaagcgttatcgcca
      L   T   P   A   T   L   S   D   A   Q   K   A   L   S   P
```

→

```
 856 gtgctggcagattggcctgcttatactgctaattttactgcggat
      V  L  A  D  W  P  A  Y  T  A  N  F  T  A  D

 901 gaagccaaagccattgcaaaagatcaattagtattagtcgcttac
      E  A  K  A  I  A  K  D  Q  L  V  L  V  A  Y

 946 gatgccgatgacaaacctgttggcgcaacctatgtgcaagcggcc
      D  A  D  D  K  P  V  G  A  T  Y  V  Q  A  A

 991 aaggtgcttgatgatctttacacacagggtgaccaagatgctgat
      K  V  L  D  D  L  Y  T  Q  G  D  Q  D  A  D

1036 gaggctacgctcggtgtggtttacgatggggcgaacattattgct
      E  A  T  L  G  V  V  Y  D  G  A  N  I  I  A

1081 tctgtgtgggcaccgactgcccaagacgtaaaactgaaagtctat
      S  V  W  A  P  T  A  Q  D  V  K  L  K  V  Y

1126 gatgccaataaaacacttaaaagcactgagaatatgaccttagat
      D  A  N  K  T  L  K  S  T  E  N  M  T  L  D

1171 agcctcactggtatctggtcctatcgcggtgatgctagccttaat
      S  L  T  G  I  W  S  Y  R  G  D  A  S  L  N

1216 cgcctctattatcgttatgcgctcagtgtgtaccatccaacgact
      R  L  Y  Y  R  Y  A  L  S  V  Y  H  P  T  T

1261 aaaaaacttgagtcgttagaaagcactgacccttattctttgaat
      K  K  L  E  S  L  E  S  T  D  P  Y  S  L  N

1306 gtgtcgagtaacggccgttttagccagtttattaatttggacgat
      V  S  S  N  G  R  F  S  Q  F  I  N  L  D  D

1351 gcagaccttaaacctgacggctgggataaccaaaccgttgctgcg
      A  D  L  K  P  D  G  W  D  N  Q  T  V  A  A

1396 attgctcacccagaagatgcggtgatctatgaggggcatattcga
      I  A  H  P  E  D  A  V  I  Y  E  G  H  I  R

1441 gactttagtattcgagatcagagcaccagtgccgcccaccgtggt
      D  F  S  I  R  D  Q  S  T  S  A  A  H  R  G

1486 aaatatttggcctttaccgagcaaggctctgcgccagtcgaacat
      K  Y  L  A  F  T  E  Q  G  S  A  P  V  E  H

1531 cttaaaaagctggttgataatggcttaactcacttccatgtgttg
      L  K  K  L  V  D  N  G  L  T  H  F  H  V  L

1576 cccgccaccgatatggcaacagtcaatgaaaatgccagtgaacgc
      P  A  T  D  M  A  T  V  N  E  N  A  S  E  R

1621 atcgaaatgaccgacacattggccaagttatgcagcaaaatcaat
      I  E  M  T  D  T  L  A  K  L  C  S  K  I  N

1666 gatacggccgatgcttgtaaaacccaaaacaagtcagcaactata
      D  T  A  D  A  C  K  T  Q  N  K  S  A  T  I

1711 gagagtattctggcgaatagcttacccggctcaagtgatgcgcag
      E  S  I  L  A  N  S  L  P  G  S  S  D  A  Q

1756 gcattagtggatgctatgcgtggtttagatggctttaactggggt
      A  L  V  D  A  M  R  G  L  D  G  F  N  W  G

1801 tacgatccgcagcatttttttgcacctgaaggcagttatgccaca
      Y  D  P  Q  H  F  F  A  P  E  G  S  Y  A  T

1846 gtggctgagggtacggcgcgagtgcttgaactgcgcgaaatgaac
      V  A  E  G  T  A  R  V  L  E  L  R  E  M  N

1891 caagccttgcataatattggtctcagagtcgtattagatgttgtg
      Q  A  L  H  N  I  G  L  R  V  V  L  D  V  V

1936 tataaccacacgagcgcttctggcgtgaatgctaacgctgtgctt
      Y  N  H  T  S  A  S  G  V  N  A  N  A  V  L

1981 gataaggtcgttcctggttattaccacagatatgaccctgtatct
      D  K  V  V  P  G  Y  Y  H  R  Y  D  P  V  S
                              ⟶
```

```
2026  ggcgcgatggaacagtcaacctgttgtgaaaataccgccactgaa
       G   A   M   E   Q   S   T   C   C   E   N   T   A   T   E
2071  catcgaatgatgggcaaactcgtcactgattctttagttttgctt
       H   R   M   M   G   K   L   V   T   D   S   L   V   L   L
2116  gccaaagaatatggctatgacggcttccgttttgatgtgatgggg
       A   K   E   Y   G   Y   D   G   F   R   F   D   V   M   G
2161  cacatgcctaagcaaggtattctcgatgcgagaaccgctgttcag
       H   M   P   K   Q   G   I   L   D   A   R   T   A   V   Q
2206  gccgtcgatccagacacttacttctacggtgaaggatggaacttt
       A   V   D   P   D   T   Y   F   Y   G   E   G   W   N   F
2251  ggtgaagttgccgataatcgtttattcgaacaggcaacccaagcc
       G   E   V   A   D   N   R   L   F   E   Q   A   T   Q   A
2296  aatatggcaggctccgaagtgggcacctttaatgacagaatacgc
       N   M   A   G   S   E   V   G   T   F   N   D   R   I   R
2341  gaggcggtgcgtggtggcgcattgtttgcgactgagtcaaaggat
       E   A   V   R   G   G   A   L   F   A   T   E   S   K   D
2386  gactacctgcgggatcaagataccttacgcttatcattggcgggt
       D   Y   L   R   D   Q   D   T   L   R   L   S   L   A   G
2431  aacttacaaaactatgtactaaaagactttaatggcaattcggct
       N   L   Q   N   Y   V   L   K   D   F   N   G   N   S   A
2476  aagggcagtagctttagctggaactcacaaccgagtgcttacgct
       K   G   S   S   F   S   W   N   S   Q   P   S   A   Y   A
2521  ctcgatcctgccgacagtattaactatgtttctaagcacgataac
       L   D   P   A   D   S   I   N   Y   V   S   K   H   D   N
2566  gaaaccctgtgggatatcctgcaatataaacatgccattggtttg
       E   T   L   W   D   I   L   Q   Y   K   H   A   I   G   L
2611  agtattgaaaatcgtgtgcgagtccaaaacatggccgccagtatt
       S   I   E   N   R   V   R   V   Q   N   M   A   A   S   I
2656  ccactcttgagccaaggggtgccgttcctgcaaatgggcggtgat
       P   L   L   S   Q   G   V   P   F   L   Q   M   G   G   D
2701  ttactgcgctcaaaatcgatggacagaaacagctatgactcgggg
       L   L   R   S   K   S   M   D   R   N   S   Y   D   S   G
2746  gactggtttaactatgtcgactttacctacaacaccaataactgg
       D   W   F   N   Y   V   D   F   T   Y   N   T   N   N   W
2791  aatgtaggtttaccactcgcgcaggataacagcagcaaatggaca
       N   V   G   L   P   L   A   Q   D   N   S   S   K   W   T
2836  cagattgcgggtatttctgctaacccaaatgcggcggcttccatg
       Q   I   A   G   I   S   A   N   P   N   A   A   A   S   M
2881  tcggatatcgaattggcctcggggggtatttaacgagtttttacgt
       S   D   I   E   L   A   S   G   V   F   N   E   F   L   R
2926  atacgccatcagagcccattattccgcctgaccgatgagcagacc
       I   R   H   Q   S   P   L   F   R   L   T   D   E   Q   T
2971  attattgagcgtgtgggattccataacgtcggtaagcgtcagcaa
       I   I   E   R   V   G   F   H   N   V   G   K   R   Q   Q
3016  cagggtgtcatcgtcatgagtctggacgacggcgttggtcttagc
       Q   G   V   I   V   M   S   L   D   D   G   V   G   L   S
3061  gatctcgacccgacagtcgatgccatcgtcgtgatgatcaacggc
       D   L   D   P   T   V   D   A   I   V   V   M   I   N   G
3106  agcgcaacggagcagtctcacacagtgccgactgcagcaggcttt
       S   A   T   E   Q   S   H   T   V   P   T   A   A   G   F
3151  agcttacacgaagtacaagcaaactctgttgatgccaaaatccgc
       S   L   H   E   V   Q   A   N   S   V   D   A   K   I   R
```

```
3196 ggcgcaagtttttagtgagggtgcgaatgaaggcacatttacggtt
      G   A   S   F   S   E   G   A   N   E   G   T   F   T   V
3241 cctgcctacagcacggctgtgtttgttaaaaaacaaagcggtgcg
      P   A   Y   S   T   A   V   F   V   K   K   Q   S   G   A
3286 cagggcttgggtctctcggcaaacgctacagttggtgcgcccgat
      Q   G   L   G   L   S   A   N   A   T   V   G   A   P   D
3331 gttgtgccctatggcagtaccgatgtatttgtccgtggcggcatg
      V   V   P   Y   G   S   T   D   V   F   V   R   G   G   M
3376 aatagctggggtgaggttgataaaattcgcctacgtcggcgctggt
      N   S   W   G   E   V   D   K   F   A   Y   V   G   A   G
3421 gaataccgtattgccatcaaactgaatgccggtgactatgatttc
      E   Y   R   I   A   I   K   L   N   A   G   D   Y   D   F
3466 aaaattgcctcggtagattggaataccgttgattttggtgccata
      K   I   A   S   V   D   W   N   T   V   D   F   G   A   I
3511 agtgctgatgtcgcgcaagttgatgaaggcgtggccgaagcgcta
      S   A   D   V   A   Q   V   D   E   G   V   A   E   A   L
3556 gcagcgaagggcgggaatctgcattttttccgcggggataacggca
      A   A   K   G   G   N   L   H   F   S   A   G   I   T   A
3601 acctatgtgttctcccttgatgcatcggataaggaccatcccgta
      T   Y   V   F   S   L   D   A   S   D   K   D   H   P   V
3646 ttaagcgtttataacgaagagccatttgtaggaacgccagtgtat
      L   S   V   Y   N   E   E   P   F   V   G   T   P   V   Y
3691 ctgcgtggtgatatgaatggttggggggactgataacgaagtcaag
      L   R   G   D   M   N   G   W   G   T   D   N   E   V   K
3736 tatatgggcggtgggatatatcgcgttgatctcgccatcacggct
      Y   M   G   G   G   I   Y   R   V   D   L   A   I   T   A
3781 ggtactaagggctttaaactcgcgtccagtgattggtctaccgta
      G   T   K   G   F   K   L   A   S   S   D   W   S   T   V
3826 gattatggtaccggtgaggccgatggtgtcgtcgtgctcaatcaa
      D   Y   G   T   G   E   A   D   G   V   V   V   L   N   Q
3871 gagaaattgctggcggtcaaaggcggcaatatgagcatcaccttc
      E   K   L   L   A   V   K   G   G   N   M   S   I   T   F
3916 gacagcgatgatacctacagctttgtgtttgatgcgtcaaatcgt
      D   S   D   D   T   Y   S   F   V   F   D   A   S   N   R
3961 aatgaacctaaactcagcgtgtataaaacggcgatgtttggtgcg
      N   E   P   K   L   S   V   Y   K   T   A   M   F   G   A
4006 aacacggtttacattcgtggcggcatgaatggttggggcacagtg
      N   T   V   Y   I   R   G   G   M   N   G   W   G   T   V
4051 gacacgttagcctatcaaggcgcttcggtttacagtgtcgatatc
      D   T   L   A   Y   Q   G   A   S   V   Y   S   V   D   I
4096 gcgttaatcgcaggtgataacgagttcaaaattgcttcatcagat
      A   L   I   A   G   D   N   E   F   K   I   A   S   S   D
4141 tggtctagtgtcgattttggtgctgtggtggggggatgaggctgtg
      W   S   S   V   D   F   G   A   V   V   G   D   E   A   V
4186 acgcttgatgggttaaaagtcttggcttcaaagggagcaaatatg
      T   L   D   G   L   K   V   L   A   S   K   G   A   N   M
4231 cattttaatgcgccagtggatgcaacttatcgctttactgtgaca
      H   F   N   A   P   V   D   A   T   Y   R   F   T   V   T
4276 ggtcccgatcctaaatcgccaaccttaactgtgactcaagttaat
      G   P   D   P   K   S   P   T   L   T   V   T   Q   V   N
4321 tag gctaacgcagcgagccgcggttaactcaccatggattcgct
      *                                              ⟶
```

```
gtgaatttaaaaccaccttaagtgcgaacttaaggtggttttttt
attcccctgagcgggagggttacaaatatttaaccgatacagagg
caaggcttaaaactgtgagcgaaacaagatatggattttctgact
tgctggttttgtcacaattgtttgtttttatttgcttaggtgtgt
ctagtggttactgcttgtttagcgttaggttaatttttatgctgc
caatgtgagctttatcaattttgatgtactgaggttttggcagaa
tgctcgttatctttttaccgagacctagcctgataaccctatgc
ataaaattctaagaatggtgaaacacacctgtcagataatgcga
ttctactatccacgacggatttgaagggaaacattaaatatgtga
atcagactttttcgcagatcagcgagtttagcgccgctgagttgc
aagggaatccccataacatagtgcgccatcaagatatgccagcgg
ctgcgtttaagatcctttgggagcgaattaagcagggtaaacctt
ggatgggtatcgttaaaaatcgcactaaaaatggcggttattact
gggtaaatgcctatgtcgcccctgtgtatgaagacggtaaaattc
acgaatatcaatctgttcgccgccaagccactccagagcagatca
aagccgccgagagtatttacagcgatattaatcagggtaagcaac
ctaaggtattgcgcaaagataggttaggtttcagtggcaagatac
tgcttgccatgctgatgtcgattatgactacggcaatgattgcca
gttattctcccattgttgctgcgattgcggggatgacggtagcgt
gcctagcatggtattacctgatgcagccgctgcaaaagctcgttt
ctatggcgacaaacattattgacgatc
```

Fig 3-6: Nucleotide sequence and deduced amino acid sequence of the pullulanase from *Shewanella arctica*. The stop codon (TAG) is marked by an asterisk. The hypothetical Shine-Dalgarno-like sequence is double underlined. The first 23 amino acids of the enzyme representing the signal sequence are indicated in bold letters. The highly conserved seven-residue motif typical of type I pullulanase (FNWGYDP) and the four regions conserved in amylolytic enzymes belonging to glycosyl hydrolase family 13 are underlined. The N-terminal amino acids determined by sequence analysis are underlined with dots.

Table 3-6: comparison of sequences of pullulanase type I from different microorganisms. All data are obtained from Genbank database.

Bacterial strain	GenBank association No.	Identity (%)	Similarity (%)	(Y/FNWGYDP) Start position and sequence	Number of total amino acids
Shewanella arctica sp. nov.	AJ877256	100	100	(598) FNWGYDPQHF	1440
Microbulbifer degradans	ZP_00315635.1	53	86	(592) FNWGYDPHHF	1432
Vibrio vulnificus	NP_936115.1	39	55	(675) FNWGYDPFHY	1202
Klebsiella pneumoniae	CAA36431.1	37	55	(563) YNWGYDPFHY	1090
Chloroflexus aurantiacus	AAN85642.1	34	55	(591) FNWGYDPFHY	1183
Bdellovibrio bacteriovorus	NP_968139.1	33	49	(357) FNWGYDPVHF	887
Bacillus cereus	YP_084052.1	30	45	(425) YNWGYDPKNF	850
Bacillus thuringiensis	YP_036824.1	30	45	(427) YNWGYDPKNF	852
Anaerobranca horikoshii	AAP45012.1	30	45	(446) YNWGYDPRLY	865
Anaerobranca gottschalkii	AAS47565.1	30	45	(446) YNWGYDPRLY	865
Clostridium perfringens	BAB81258.1	29	47	(213) YNWGYDPVNY	655
Exiguobacterium sp.	ZP_00182722.2	29	44	(548) YNWGYDPKNY	970
Lactobacillus acidophilus	YP_194553.1	29	46	(594) YNWGYDPKNY	1185
Geobacillus kaustophilus	YP_148680.1	29	44	(293) YNWGYDPLHL	718
Thermus thermophilus	BAB62095.1	29	44	(293) YNWGYDPLHL	718
Bacillus subtilis	NP_390871.1	29	44	(288) YNWGYDPLHF	718
Bacteroides thetaiotaomicron	AAC44685.1	29	47	(234) YNWGYDPQNY	668
Fervidobacterium pennivorans	AF096862_1	29	43	(424) YNWGYDPYLF	849
Francisella tularensis	YP_169457.1	26	41	(630) FNWGYDPQNY	1070
Bacillus halodurans	BAB06983.1	26	44	(296) YNWGYDPVHY	717
Streptococcus pyogenes	ZP_00365686.1	26	40	(566) YNWGYDPQHY	1114
Oceanobacillus iheyensis	NP_691326.1	26	42	(281) YNWGYDPLFF	721
Thermotoga maritime	NP_229641.1	26	47	(416) YNWGYDPYLF	843
Streptococcus suis	ZP_00332302.1	25	41	(273) YNWGYDPMQY	683
Streptococcus pneumoniae	AF217414_1	23	37	(670) YNWGYDPQNY	1287
Clostridium acetobutylicum	NP_349286.1	23	49	(286) YNWGYDPQNY	720
Mycoplasma mobile	YP_016097.1	20	46	(301) YNWGYDPLQW	722

Table 3-7: The sequences of four conserved regions of type I pullulunases *

Source	Starting position and amino acid sequence for conserved region		
	Y/FNWGYDP	I	II
Shewanells arctica **sp. nov.**	598 FNWGYDP	636 GLRVVLDVVYNHT	713 DGFRFDVMGH
Microbulbifer degradans	592 FNWGYDP	630 GLRVVLDVVYNHT	707 DGFRFDIMGH
Klebsiella pneumoniae	563 YNWGYDP	602 GMNVIMDVVYNHT	680 DGFRFDLMGY
Vibrio vulnificus	675 FNWGYDP	741 GMNVVMDVVYNHT	791 DAFRWDLMGH
Chloroflexus aurantiacus	591 FNWGYDP	629 GLRVVMDVVYNHT	705 DGFRFDLMGH
Bdellovibrio bacteriovorus	357 FNWGYDP	395 NLRVVQDVVFNHT	471 DAFRFDLMSF
Streptococcus suis	280 YNWGYDP	629 GLRVVLDVVYNHT	393 DGFRFDLMGI
	III	IV	
Shewanells arctica **sp. nov.**	745 YGEGWNF	844 PADSINYVSKHDNETLWD	
Microbulbifer degradans	739 YGEGWNW	831 PADIINYVSKHDNETLWD	
Klebsiella pneumoniae	712 FGEGWDS	831 PTEVVNYVSKHDNQTLWD	
Vibrio vulnificus	823 YGEGWNF	948 AWEIQNYVSKHDNQTLWD	
Chloroflexus aurantiacus	745 YGEGWNF	867 PQEQIVYVSAHDNETLFD	
Bdellovibrio bacteriovorus	511 YGEGWAF	639 ALETVSYVSAHDGYTLWD	
Streptococcus suis	712 FGEGWDS	497 PQQSLNYIECHDNATAFD	

* The sequences are driven from GenBank database and have the following accession no.: *Microbulbifer degradans* = ZP_00315635.1, *Klebsiella pneumoniae* = CAA36431.1, *Vibrio vulnificus* = NP_936115.1, *Chloroflexus aurantiacus* = AAN85642.1, *Bdellovibrio bacteriovorus* = NP_968139.1 and *Streptococcus suis* = ZP_00332302.1. The underlined amino acid residues are the residues identified by Nakajima et al. (1986) (Nakajima *et al.*, 1986) as being highly conserved.

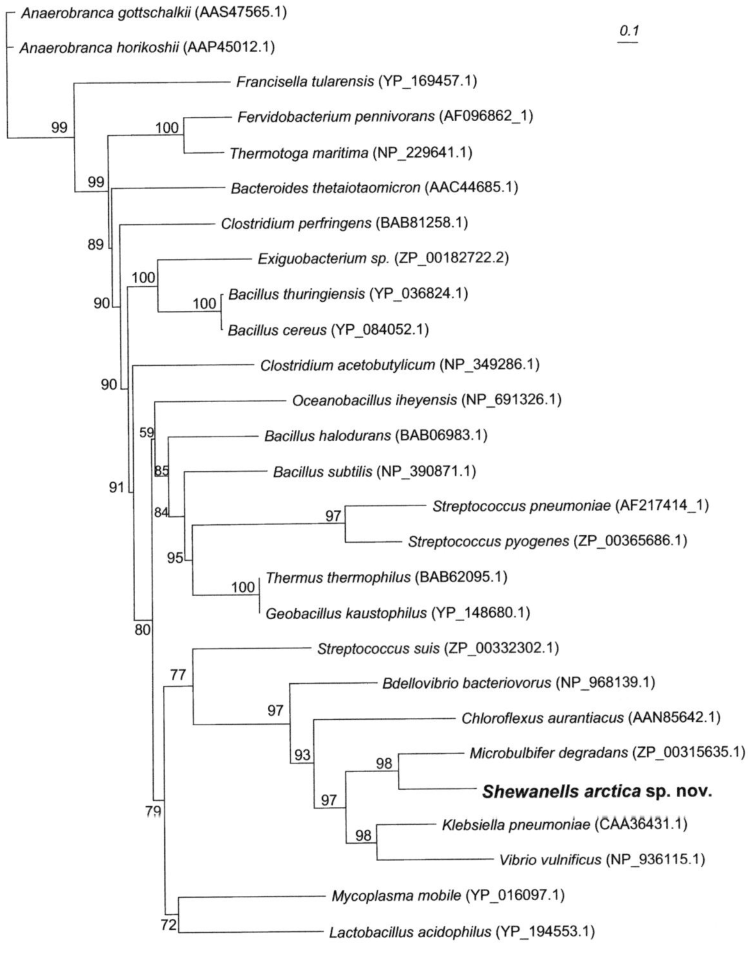

Fig. 3-7: Phylogenetic dendrogram based on amino acids sequences of the new pullulanase from *Shewanella arctica* and the most related pullulanase from different bacterial strains. Performed by the neighbour-joining method using software from PHYLIP, version 3.57c; the PRODIST program with Kimura-2 facter was used to compute the pairwise evolutionary distances for the above aligned sequences , the topology of the phylogenetic tree was evaluated by performing a bootstrap (algorithm version 3.6 b) with 1000 bootstrapped trials. The tree was draw using Tree View 32 software.

3.3.3 Purification of the native pullulanase

Analysis of the pullulanase produced from *Shewanella arctica* sp nov. cells shed light on the true size of the enzyme as it occurs in the natural producer organism. The culture supernatant, concentrated by ultrafiltration of a 20 l fermentation of *Shewanella arctica* sp nov. showed a very low pullulanase activity, which could be only seen on red-pullulan screening agar plates. Therefore, cells of *Shewanella arctica* sp nov. containing cell-associated enzyme were a better source for the wild-type pullulanase. The specific activity of the pullulanase in the cell crude extract was 0.055 U/mg. Purification of the pullulanase after, Q-Sepharose chromatography, phenyl-sepharose chromatography, hydroxylapatite chromatography, gel filtration on Superdex S-200 and Mono-Q chromatography resulted in a 54 fold increase in specific activity as shown in Table 3-8. All purification steps were carried out at room temperature. The activity at this stage was enough to perform characterization of the native enzyme. Proteins from the purification steps were separated by 8 % SDS-PAGE. The enzyme migrates as a single band with an apparent molecular mass of 155 kDa. This matched the molecular mass of the single active band detected on the zymogram pattern (SDS-8 % PAGE activity gel of the purified native enzyme) (Fig.3-8). This is also in agreement with the predicted molecular mass detected from the amino acid sequence (Fig. 3-6). This indicates that the active native enzyme is a monomer.

Table 3-8: Purification of the native pullulunase from *Shewanella arctica*[$]

Purification step	Total protein (mg)	Total activity* (U)	Specific Activity (U/mg)	Recovery (%)	Purification (Fold)
Crude extract	1225	68.24	0.055	100	1
Q-Sepharose	235	58.75	0.25	86	4.5
Phenyl-Sepharose	17.1	10	0.58	14.7	10.5
Hydrohylapatite	1.02	2.076	2.035	3	37
Superdex S-200	0.48	1.248	2.6	1.8	47
Mono Q	0.246	0.738	3.0	1	54.5

[$] After the growth of *Shewanella arctica* at 10 °C, a 5L culture was centrifuged (80 g of wet weight cells), cells were washed 3 times with PBS buffer, and resuspended with Tris-NaCl-Triton X-100 buffer (pH 8). Cells were disrupted with French press, centrifuged and the supernatant was used for purification.

* One unit of pullulanase is defined as the amount of enzyme that releases 1 µmol of reducing sugar per min under assay conditions specified. Maltotiose was used as a standard.

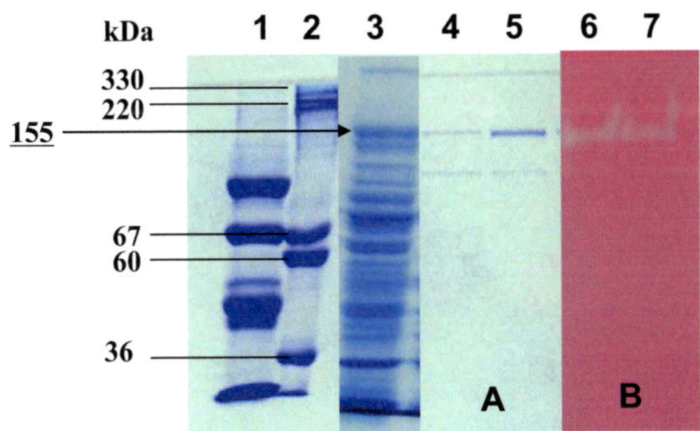

Fig. 3-8: Gel electrophoretic analysis (8 % SDS-PAGE) and zymogram of the purified native pullulanase from *Shewanella arctica*. (A) Purification of native pullulanase. Lane 1 and 2, marker proteins; lane 3, crude extract (25 µg); lane 4, mono Q pool (13 µg); lane 5, mono-Q pool (25 µg). (B) Zymogram of the purified native pullulansae. Lane 6 and 7, mono-Q pool (25 µg each). Arrow indicates the position of pullulanase.

Edman degradation analysis of the purified native pullulanase was also carried out, to reveale the N-terminal sequence. These 18 amino acids found in the N-terminal sequence (CGGSDSSEPGKVLLTCDV) driven from the analysis of the purified native pullulanase, was also found in the N-terminal sequence (amino acid residues number 24 to 41 downstream) of the recombinant pullulanase gene as shown in Fig. 3-6.

3.3.4 Purification of the recombinant pullulanase after expression in *E. coli* XLOLR

Expression of the recombinant pullulanase in *E. coli* XLOLR was studied with and without induction (1 mM IPTG), at 30 °C and 37 °C and at different incubation times (6–55 hours). Optimal expression conditions for the recombinant enzyme production was determined by measuring the enzyme activity (U/mg) of the crude extract from the *E.coli* XLOLR clone (as described in section 2.4.4). The highest specific activity (0.54 U/mg) was reached after 36 hour growth on LB-kanamycin medium, without induction with IPTG at 30 °C (Fig. 3-9).

Purification of recombinant pullulanase was carried out from the crude extract from *E.coli* XLOLR cells grown at optimal expression conditions. The results of the purification procedure are summarized in Table 3-9. After Q-sepharose, hydroxylapatite, Superdex S-200 gel filtration and Mono Q chromatography the recombinant pullulanase was purified 3.24, 7.08, 107.7 and 170.5- Fold, with a specific activity of 1.27, 2.77, 42.14, and 66.6 U/mg, and yield of 26.5, 11.3, 0.8, and 0.2 % respectively. Proteins from the purification steps were separated using SDS-PAGE-8 %.

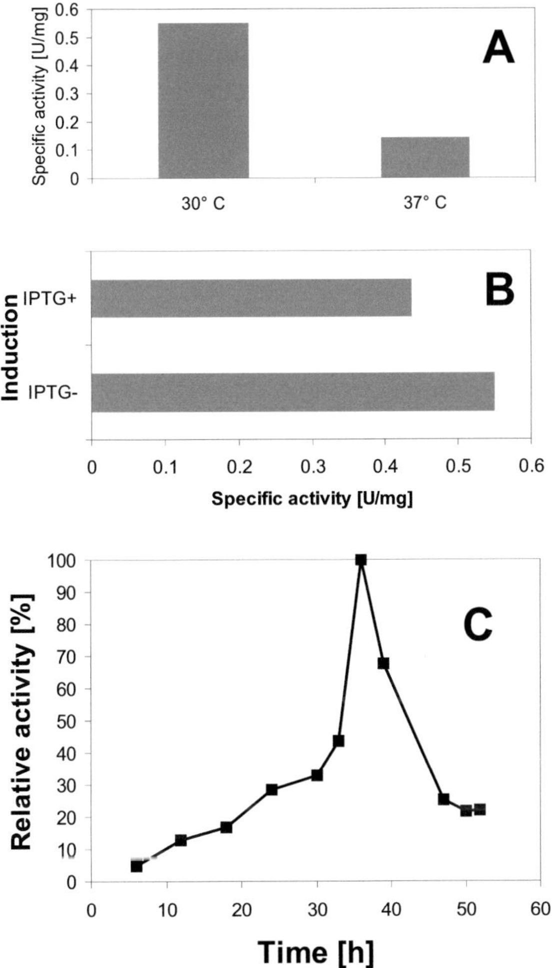

Fig. 3-9: Factors influencing the expression of recombinant pullulanase from *E. coli* XLOLR clone. (A) In order to determine the optimum temperature of the expression of the enzyme produced by *E. coli* XLOLR clone, the cells were grown without IPTG at 30 °C and 37 °C for 36 h. and activity was measured in cell extracts. (B) The influence of the presence and absence of IPTG (1 mM) on enzyme production was investigated at 30 °C for 36h. (C) For determination of the optimum incubation time for enzyme production, the cells were incubated at 30 °C without IPTG for 55 h. The enzyme activity was tested in samples taken at various periods of time (0 to 55 h). LB-Kanamycin liquid medium was used as culture medium (section 2.3.4).

Table 3-9: Purification of the recombinant pullulanase of *Shewanella arctica* after expression in *E. coli* XLOLR[$]

Purification step	Total protein (mg)	Total activity* (U)	Specific Activity (U/mg)	Recovery (%)	Purification (Fold)
Crude extract	73000	28600	0.39	100	1
Q-Sepharose	6000	7600	1.26	26.5	3.2
Hydrohylapatite	1170	3240	2.7	11.3	7
Superdex S-200	5.6	236	42	0.8	107
Mono Q	0.96	64	66.6	0.2	170

[$] After the aerobic growth of *E. coli* XLOLR at 30 °C in 16 L culture cells were centrifuged (40 g of wet weight cells), resuspended with Tris-NaCl-Triton X-100 buffer (pH 7), disrupted with French press and the supernatant was used for purification.

* One unit of pullulanase is defined as the amount of enzyme that releases 1 μmol of reducing sugar per min under assay conditions specified. Maltotiose was used as a standard.

The samples after hydroxylapatite chromatography revealed two protein bands with apparent molecular masses of 85 and 70 kDa. The zymogram staining showed that both bands posess pullulanase activity by forming clear zones on dyed pullulan (Fig. 3-10). The total molecular mass of both bands masses (155 kDa) is similar to the size of the pure native pullulanase (Fig. 3-9) and to the calculated molecular mass from the predicted amino acids sequence (Fig. 3-6). After gel filtration and Mono-Q chromatography only the 70 kDa band was present (Fig. 3-10). Zymogram staining, however, shows that the single polypeptide was able to degrade the dye-pullulan showing an active band (Fig. 3-10).

Since SDS-PAGE and the zymogram staining analysis of the native purified pullulanase showed that the full-length (155 kDa) native enzyme is a monomer, the two active protein bands (70 and 85 kDa) indicate that the recombinant pullulanase (155 kDa) is probably hydrolyzed by endogenous protease, leading to the formation of two proteins with the sizes of 70 kDa and 85 kDa

3.3.5 Biochemical properties of native and recombinant pullulanases

Native and recombinant pullulanases were both active between 4 °C and 60 °C. The temperature optimum of both pullulanases was 45 °C and a rapid decrease in pullulanase activity was observed above 45 °C and 50 °C for the native and recombinant enzyme, respectively (Fig. 3-11A). Native and recombinant pullulanases showed activity over a broad pH range (native: pH 4-9, Recombinant: pH 6-8), with the same optimum at pH 7 (Fig. 3-11B). The influence of temperature on the stability of the recombinant pullulanase was examined by measuring the enzymatic activity after incubating the recombinant enzyme at low temperatures (4, -20, and -80 °C) (Fig. 3-12B) and at high temperatures (40, 50 and 65 °C) (Fig. 3-12A). After three days of incubation at -80 °C the enzyme lost 55 % of its initial activity. The 55 % lost of initial activity was measured after four and ten days of incubation at 4 °C and -80 °C respectively. The enzyme is more stable at -20 °C rather than 4 °C or -80 °C. The recombinant enzyme is fully stable at 40 °C for 3 hours and exhibits a half-life of 44 min at 50 °C. A complete loss of activity was observed at 65 °C after 30 min.

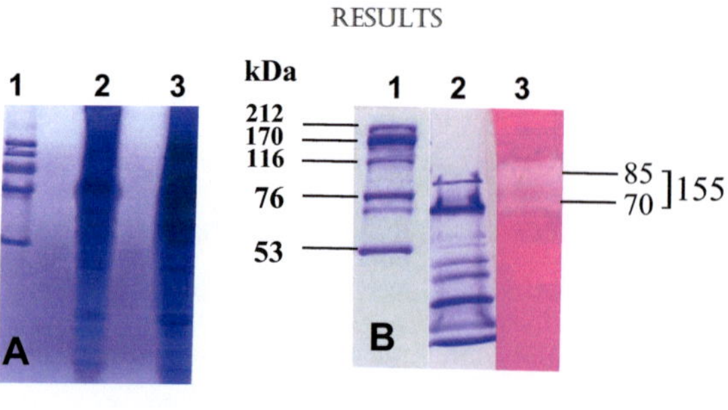

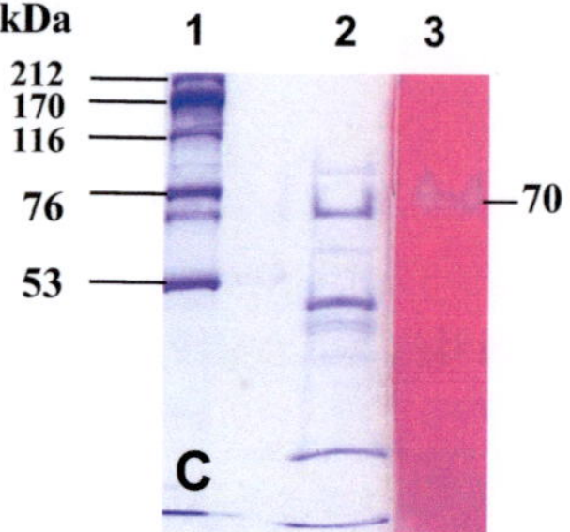

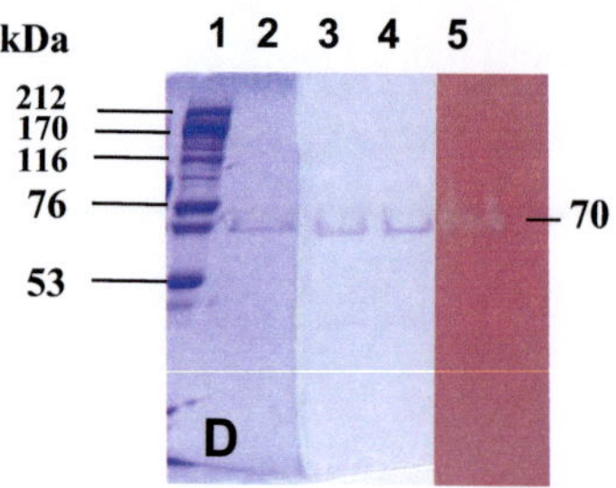

Fig. 3-10: Gel electrophoretic analysis (8 %-SDS-PAGE) and zymogram of the partial purified recombinant pullulanase. (A) Purification of recombinant pullulanase. Lane 1, marker proteins; lane 2, Q-Sepharose pool (20 µg); lane 3, crude extract (20 µg). (B) Purification of recombinant pullulanase. Lane1, marker proteins; lane 2, hydroxylapatite pool (20 µg); lane 3, zymogram of hydroxylapatite pool (20 µg) showing the 2 active pullulanase bands with a molecular size of 85 and 70 kDa. (C) Purification of recombinant pullulanase. Lane1, marker proteins; lane 2, gel filtration pool (20 µg); lane 3, zymogram of gel filtration pool (20 µg) showing one active band (70 kDa). (D) Purification of recombinant pullulanase. Lane1, marker proteins; lane 2, 3 and 4, mono Q pool (20 µg each); lane 5, zymogram of mono Q pool (20 µg) showing one active band (70 kDa).

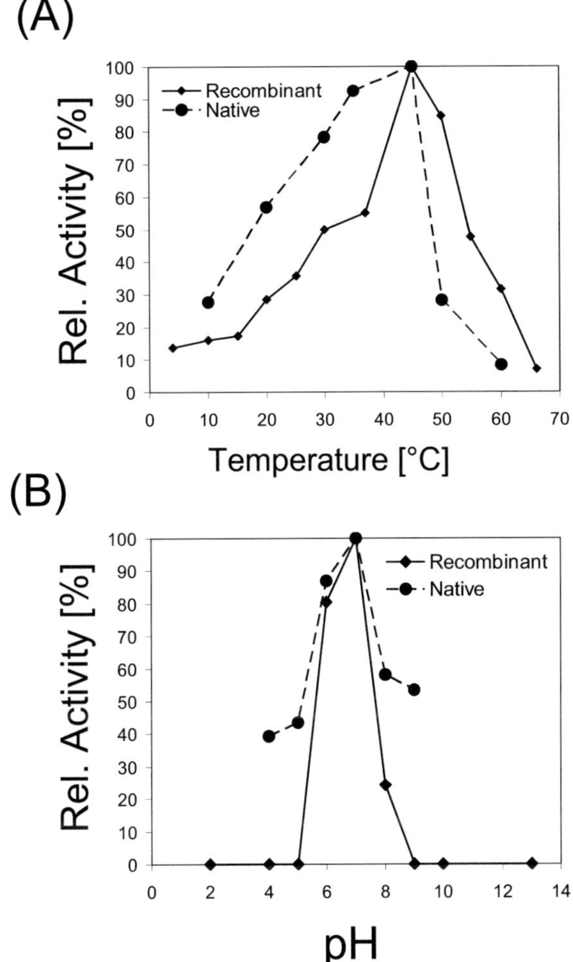

Fig. 3-11: Influence of temperature (A) and pH (B) on the activity of the native and recombinant pullulanase from *Shewanella arctica*. For the determination of optimum temperature the native (3 U/mg) and the recombinant (2.7 U/mg) enzymes were incubated in Tris-HCl buffer (pH 7) for 30 min at different temperatures. For the determination of pH optimum the native and recombinant enzymes were incubated in sodium acetate buffer (pH 2 to 5) and Tris-HCl buffer (pH 6 to 13) at 45 °C for 30 min at different pH values. (♦) native enzyme and (●) recombinant enzyme.

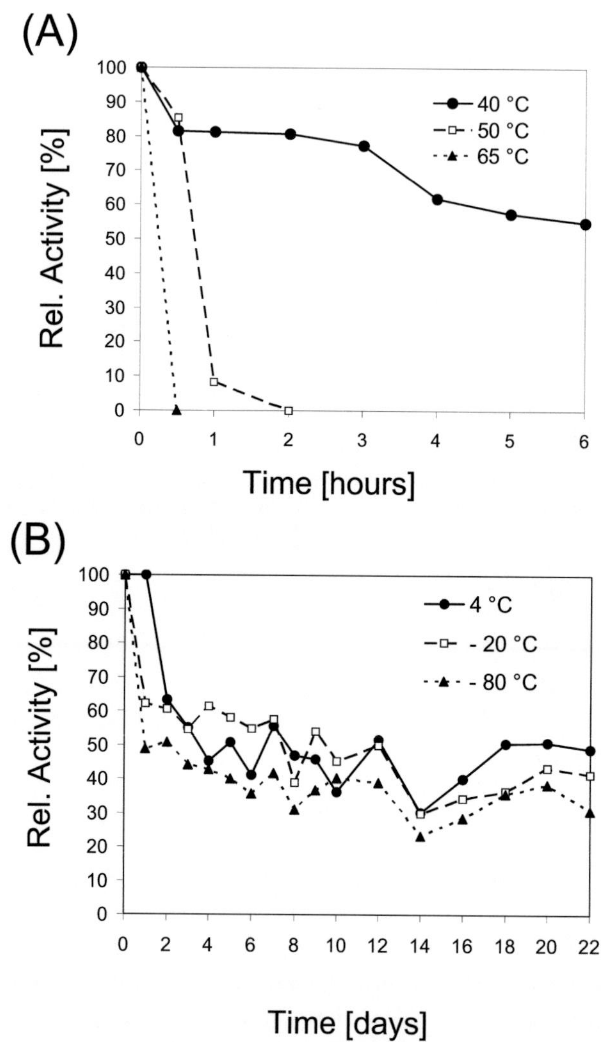

Fig. 3-12: Thermostability of the recombinant pullulanase from *Shewanella arctica*. The enzyme (2.7 U/mg) was incubated at (A) 40 °C (●), 50 °C (□), and 65 °C (▲). (B) 4 °C (●), -20 °C (□), and -80 °C (▲). Samples were withdrawn and tested for pullulanase activity.

3.3.6 Effect of metal ions and other reagents on native and recombinant pullulanase activity

The effect of metal ions and different reagents on *Shewanella arctica* sp. nov. native and recombinant pullulanases is shown in Table 3-13. Divalent cations affected the activity of both native and recombinant enzyme as follows: the native pullulanase was inhibited by 1mM of Mn^{+2}, Zn^{+2}, Cu^{+2} (100 % inhibition) and Fe^{+2} (100 % inhibition), whereas, Ca^{+2}, Co^{+2} and Ni^{+2} had no effect (Fig. 3-13A). The recombinant pullulanase activity was inhibited by 1mM Mn^{+2}, Fe^{+2}, Cu^{+2}, Ni^{+2}, Mg^{+2} and Zn^{+2}, no effect was measured after incubation with 1mM Fe^{+3} and Ca^{+2},while an increase in activity was measured after incubation with 1mM Co^{+2} (Fig. 3-13A).This is in contrast to reported pullulanase requiring Ca^{+2} ions for full activity. Complete inhibition of both native and recombinant enzyme activity as measured by incubating the enzyme with EDTA suggests that this reagent chelate a possible divalent cation(s) required for the activity of the enzyme. The activity of both native and recombinant enzymes was not affected by α-, β-, and γ-Cyclodextrins, which are generally known as possible competitive inhibitors of archaeal pullulanases. The activity of both enzymes was also not affected by N-bromosuccinimide, which oxidizes typtophan residues. Both recombinant and native pullulanase activity were inhibited by SDS (sodium dodecysulfate), urea, DTT (dithiothreitol), guanidine-HCl and iodocetamide (Fig. 3-13B).

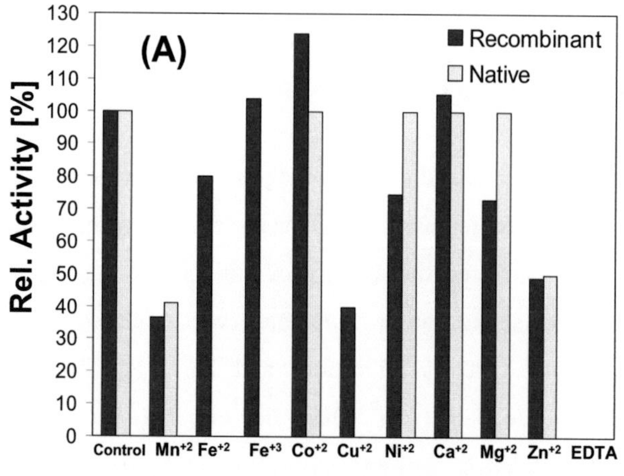

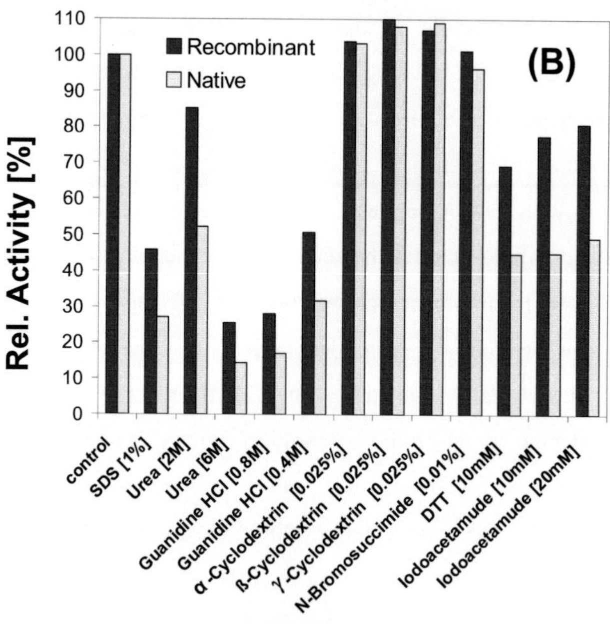

Fig. 3-13: Influence of different metal ions, EDTA (A) and reagents (B) on native and recombinant pullulanase activity. The effect of various substances on the enzyme activity was examined after incubation of native (3 U/mg) and recombinant (2.7 U/mg) enzymes with metal ions (I mM final concentration) and reagents (at various concentrations) at room temperature for 60 min. Samples were withdrawn and tested for pullulanase activity at 45 °C for 30 min and pH 7. The enzyme activity without metal ions and reagents was considered as

3.3.7 Substrate specificity and kinetic properties

A number of different substrates were incubated with the native and recombinant pullulanase in order to determine the substrate specificity. As shown in Fig. 3-14, native and recombinant pullulanase preferentially hydrolyzed pulluan. Amylopectin and starch were hydrolyzed with a high velocity, compared to amylose, which was cleaved at very low rate. This indicates that the enzyme is able to hydrolyze α-1,6 glycosyl linkages.

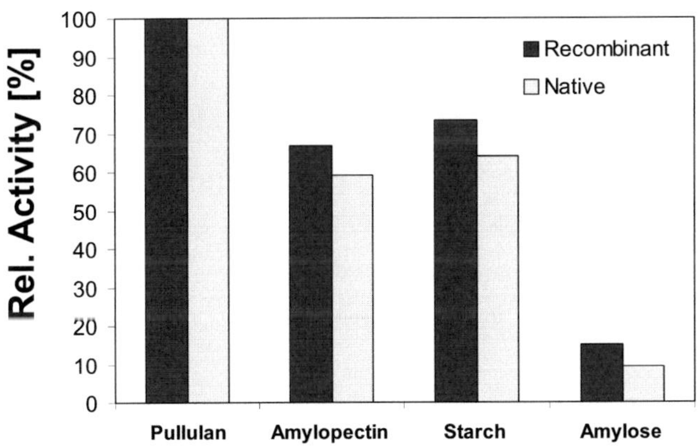

Fig. 3-14: Hydrolysis of various substrates by native and recombinant pullulanase. Substrates were dissolved in 50 mM Tris-HCl buffer (pH 7) at a final concentration of 0.5 % (w/v), 50 µl enzyme mixture (3 U/mg native and 2.7 U/mg recombinant) was added and incubation was performed at 45 °C for 30 min. Pullulan hydrolysis rate was chosen as 100 %

In order to analyze the product after pulullan was incubated 1 hour at 45 °C with the recombinant pullulanase, the products were identified by HPLC analysis (Fig. 3-15). The hydrolysis pattern after the pullulanase action on pullulan revealed the complete conversion of pullulan to maltotriose. In order to confirm that the hydrolysis product from pullulan was maltotriose (possessing two α-1,4 glycosyl linkages) and not panose or isopanose (possessing α-1,4 and α-1,6 glycosyl linkages), the product of pullulan hydrolysis was incubated with α-glycosidase from yeast. The formation of glucose as the main product confirmed the formation of maltotriose from pullulan. These results confirm together with the low affinity toward amylose that the pullulanase from *Shewanella arctica* is pullulanase type I able to hydrolyze only α-1,6 glycosyl bounds in polymeric substrates

The apparent kinetic parameters according to Michaelis-Menten- with different concentrations of pullulan (0.1-0.7 % (w/v)) were determined at pH 7 at 45 °C. The K_m value for pullulanase is 0.1768 % (w/v) and, V_{max} value is 6.77 µmol/min as shown in Fig. 3-16.

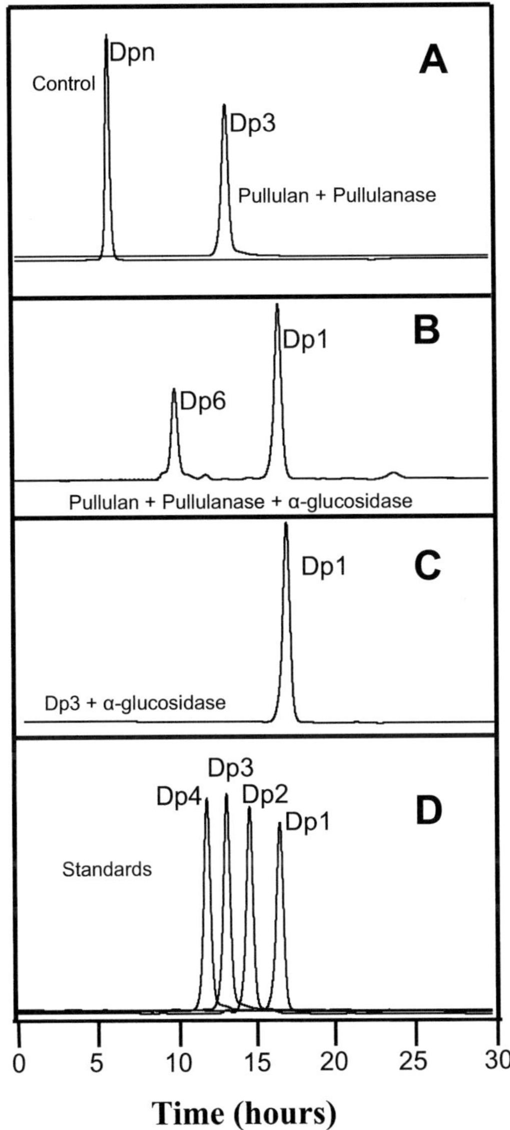

Fig. 3-15: HPLC analysis of the hydrolysis products formed after incubation of recombinant pullulanase (2.7 U/mg) in with 0.5 % (w/v) pullulan. Samples were incubated at 45 °C for 60 min and analyzed on an Aminex HPX 42-A. (A) Hydrolysis after incubation with the recombinant pullulanase. (B) The products formed after the hydrolysis products of pullulan were incubated with commercial α- glucosidase and then analyzed by HPLC. (C) Maltotriose = Dp3 after incubation with commercial α- glucosidase. (D) Standards, Dp, degree of polymerization Dp1 = glucose, Dp2 = maltose, Dp3 = maltotriose; etc.

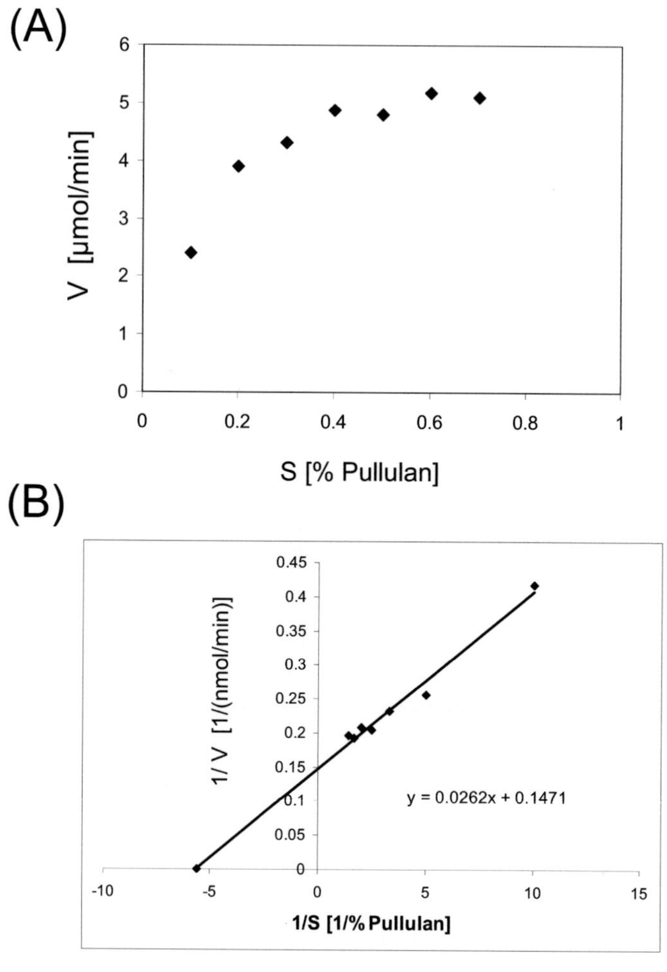

Fig. 3-16: Michaelis-Menten-Kinetics of recombinant pullulanase. (A) Effect of substrate concentration on the velocity of enzyme. (B) Double-reciprocal or Linewearer-Burk plot. In order to calculate the k_m and V_{max}, the enzyme activity was tested after Bernfeld (1955)(Bernfeld, 1955) at 45 °C, pH 7 and for 30 min with different concentrations of pullulan (0.1 to 0.7 % [w/v]). For each reaction 50 µl (2.7 U/mg) enzyme solution was used. One unit of pullulanase is defined as the amount of enzyme that releases 1 µmol of reducing sugar per min under assay conditions specified. Maltose was used as a standard.

3.4 Cloning, sequencing, purification and characterization of a subtilisin-like serine protease from *Shewanella arctica*. sp. nov.

3.4.1 Cloning and sequencing of the 6-kb insert encoding protease from *Shewanella arctica* sp. nov..

Forty thousand phagemid clones obtained from a λZAP (Stratagene) genome bank were screened for protease activity at 30 °C for 3-4 day as described in material and methods section (2.3.3). Seven clones showed clearing zones on the AZO-casein plates. PCR analysis technique revealed that all clones harbored the same gene, although the insert varied in size. A 6-kb insert was chosen for further analysis. The insert was sequenced by using primers walking technique. Four forward and four reverse walking with 7 constructed primers (Fig. 3-17) were necessary to determine the complete sequence of the insert. M13 forward and M13 reverse standard primers were used to start the primer walking. Complete sequence revealed that the G+C content was 47.42 % and that two open reading frame (ORF) were present as shown in Fig. 3-18. ORF1 (1.011 kb) encodes a protein of 336 amino acids with a predicted molecular mass of 37.2 kDa. ORF2 (1.935 kb) encodes a protein of 644 amino acids with a predicted molecular mass of 67.4 kDa. The BLAST search (NCBI database) showed that ORF1 is 88 % identical to a peptidase from *Shewanella onedensis* and 62 % identical to a putative protease-like from *Vibrio cholerae*, whereas, ORF2 is 65 % identical to an alkaline subtilisin-like serine protease from

Pseudoalteromonas sp.. Accordingly, ORF2/gene could be considered as a new subtilisin-like serine protease.

Table 3-17: Oligonucleotides used for the primer walking* of the 6-kb insert encoding different protease genes

walk	Primer direction F = forward R = reverse	Primer position (bp)	Sequences (5′ to 3′)
1	F	456	GCC GTT TTA TCA GGA GAT G
	R	719	TGG GCA GGT TGA AAA TAC G
2	F	1210	GCC TTT GAG AGT TCC ACT G
	R	1331	CGC AAG CAG TGG CTA TGA C
3	F	1722	TGA CCC TCG GAA ACT TTT G
	R	2176	ATG CGG ATA TTT ATG TGA G
4	R	3024	CTG CTG CCA CGA GTA TGA G

* M13 forward and M13 reverse standard primers were used to start the primer walking.

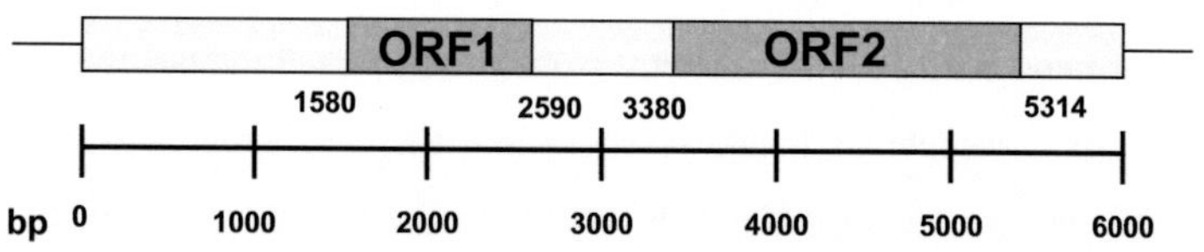

Fig. 3-18: Map of the 6-kb insert encoding the protease from *Shewanella arctica*. Whole insert of 6 kb has G+C content of 46.82 % and two open reading frames (ORFs) (marked in gray). The ORF1 start codon ATG is at 1580 bp position, 1.011kb and encodes a protein of 336 amino acid with 37.2 kDa predicted molecular mass. The ORF2 start codon ATG is at 3380 bp which includes also a signal sequence, 1.935 kb and encodes a protein of 644 amino acids with 67.4 kDa predicted molecular mass.

3.4.2 Subcloning of protease genes and their expression.

The ORF1 and ORF2 (Fig. 3-18) each with the signal amino acids sequences were subcloned into expression vector pET Blue-1 and expressed in *E. coli* Tuner (DE3) plLacl as described in section 2.3.5. F1580:ATGGAGCTGTTGTGTCCAGCAGGA and R2590:TTATTGC CAATCACGCTCGTAGGC forward and reverse primers were used for theamplification of ORF1. F3380:ATGTCTCGAAGTCCGTGGAGAC ATAAA and R5314:TTAGTAGCTTGCGCTCAGAAC the forward and reverse primers were used for the amplification of ORF2/gene. Protease activity could be detected as clearing zones around the active subclones after 2-3 days of growth on LB/carbenicillin/chloramphenical agar plates containing 5 % (w/v) AZO-casein at 30 °C. Clones having protease activity were detected only from the subcloned protease ORF2/gene and were used for further studies

3.4.3 Protease gene (ORF2) analysis and phylogenetic position.

The protease gene (ORF2) (Fig. 3-19) form *Shewanella arctica* sp. nov. has ATG (M) as start codon, TAG as stop codon, is 1.935 kb long, has a G+C content of 48.78 % and encodes an enzyme with predicted molecular mass of 67.4 kDa (644 aa). In addition, a signal sequence consisting of 38 amino acids with predicted cleavage taking place between an Alanine residue and an Asparagine residue was found. The Shine-Dalgarno sequence (AGGAGG), however was not found in the predicted amino acids sequence

```
          ..........aatgtcgcttgataaatcaccacttttacaaacg

          ttgttactttaatagactgagagttattattttgacgcttatttg

          tacaactaaaaaatcattccagtaaaaattcactcaggcccaaat

          tgtaaatcaagtgttacattttgatattgtgaatgaggtaagcct

          taagcattaaactgggcaggttgaaaatacgttctacattggatt

     1    atgtctcgaagtccgtggagacataaaaagaaaatggagttcaac
          M   S   R   S   P   W   R   H   K   K   K   M   E   F   N

    46    atgcataagaaacatttaatagcagtcgcagtcgcaacaggactt
          M   H   K   K   H   L   I   A   V   A   V   A   T   G   L

    91    gcttacttccctgttaacgctaatgaataccaagcgactatggta
          A   Y   F   P   V   N   A   N   E   Y   Q   A   T   M   V

   136    agtgtcccacaatctaaagccatcaaagatacttacatcgttgta
          S   V   P   Q   S   K   A   I   K   D   T   Y   I   V   V

   181    ttcaataccccaagtgttcttaatctaagtaataacaacaccata
          F   N   T   P   S   V   L   N   L   S   N   N   N   T   I

   226    gctgaattcgcggttcaacaagccgagagtttagtcaatcaatat
          A   E   F   A   V   Q   Q   A   E   S   L   V   N   Q   Y

   271    gatgtcagagtgatgaaaaactttggcaatgtgctcaacggtgta
          D   V   R   V   M   K   N   F   G   N   V   L   N   G   V

   316    ctcatcaatgccagtgcccaacaagttaaagcactgcttaaagat
          L   I   N   A   S   A   Q   Q   V   K   A   L   L   K   D

   361    ccaaacgtgaagtacgtagaacaagatcaagtgatgtcagtaacg
          P   N   V   K   Y   V   E   Q   D   Q   V   M   S   V   T

   406    cccatgatggaagccaatgcggaccaaccgagtccgacctggggc
          P   M   M   E   A   N   A   D   Q   P   S   P   T   W   G

   451    atagacagaatcgatcaacgcaacttgccattggataacaactat
          I   D   R   I   D   Q   R   N   L   P   L   D   N   N   Y

   496    cacacggattacgatggatcaggtgtgaccgcctttgttattgat
          H   T   D   Y   D   G   S   G   V   T   A   F   V   I   **D**

   541    actggggtgctaaatacacacaatgagttcggcggccgcgcaagc
          T   G   V   L   N   T   H   N   E   F   G   G   R   A   S

   586    agtggctatgactttatcgataatgattacgatacgactgactgt
          S   G   Y   D   F   I   D   N   D   Y   D   T   T   D   C

   631    aacggtcatggtacccatgtggcggggacgattggcggctcaacc
          N   G   **H**   G   T   H   V   A   G   T   I   G   G   S   T

   676    tacggtgtcgcgaaaaacgtcaatgtggtgggcgtcagagtgctt
          Y   G   V   A   K   N   V   N   V   V   G   V   R   V   L

   721    aactgttcaggttctggcagtaactctggcgtgattgcaggata
          N   C   S   G   S   G   S   N   S   G   V   I   A   G   I

   766    aactgggtgaaaaacaatgcttctggccccgctgtcgcgaacatg
          N   W   V   K   N   N   A   S   G   P   A   V   A   N   M

   811    agtttaggggggcggcgcctcccaagccacggatgatgccgtcaat
          S   L   G   G   G   A   S   Q   A   T   D   D   A   V   N

   856    gccgctgttgccgcagggatcaccttcgtcgtcgcagccggcaat
          A   A   V   A   A   G   I   T   F   V   V   A   A   G   **N**

   901    gacaatagtaatgcctgtaattactccccagctcgtgccgcagat
          D   N   S   N   A   C   N   Y   S   P   A   R   A   A   D
                                                        ———————→
```

```
 946 gccatcactgtcggttcaaccaccagtaacgattcccgctcgagt
      A  I  T  V  G  S  T  T  S  N  D  S  R  S  S
 991 ttttctaactacgggacttgccttgatatctatgcgcccggttcg
      F  S  N  Y  G  T  C  L  D  I  Y  A  P  G  S
1036 agcataacttcctcttggtatacctcaaattcagcgactaatacc
      S  I  T  S  S  W  Y  T  S  N  S  A  T  N  T
1081 attagtggcacctcaatggcttccccccatgtggcaggcgtcgcg
      I  S  G  T  S  M  A  S  P  H  V  A  G  V  A
1126 gcattatacttagatgaaaaccctaacctctcccccgcacaggtg
      A  L  Y  L  D  E  N  P  N  L  S  P  A  Q  V
1171 acgaacttactcaagacgcgcgccactgcggacaaagtcacagat
      T  N  L  L  K  T  R  A  T  A  D  K  V  T  D
1216 gctaagacaggctcaccgaataagttactgttttcacttgcaaat
      A  K  T  G  S  P  N  K  L  L  F  S  L  A  N
1261 gatgatggtggctgtggcaatgattgccccgttgacgagactcag
      D  D  G  G  C  G  N  D  C  P  V  D  E  T  Q
1306 ctgcaaaataatgtgggtattgcgatcagtggagccacaggttca
      L  Q  N  N  V  G  I  A  I  S  G  A  T  G  S
1351 gcgacttattactatcgatgtccccgcaaatgcggcaagttta
      A  T  Y  Y  Y  I  D  V  P  A  N  A  A  S  L
1396 ggcatcaacctcgcggggggctctggcgatgcggatatttatgtg
      G  I  N  L  A  G  G  S  G  D  A  D  I  Y  V
1441 agccaaggacaaaaaccgactacgaccagctatcaatgccgccca
      S  Q  G  Q  K  P  T  T  T  S  Y  Q  C  R  P
1486 tatcaaaatggcaacaatgagagctgtaatttcactgcacctacg
      Y  Q  N  G  N  N  E  S  C  N  F  T  A  P  T
1531 gcgggtcgttggtacgtgatggttcaaggctatagcaattatgcc
      A  G  R  W  Y  V  M  V  Q  G  Y  S  N  Y  A
1576 aacgcccagctgacagctagctacaacctcaatggcggcggaaat
      N  A  Q  L  T  A  S  Y  N  L  N  G  G  G  N
1621 tgtaccgatgcgaactgcttaaccaatggcgtaccgtcacgaat
      C  T  D  A  N  C  L  T  N  G  V  P  V  T  N
1666 ttaagcggagcaacgggaacggaagccctgtataaaatcgtcgtc
      L  S  G  A  T  G  T  E  A  L  Y  K  I  V  V
1711 ccagcgaatagccaactcagtattaccaccagtggcgggactgga
      P  A  N  S  Q  L  S  I  T  T  S  G  G  T  G
1756 gacgtggatctgtatgtcaaagcagggactgtcccaacgactacc
      D  V  D  L  Y  V  K  A  G  T  V  P  T  T  T
1801 agctatgattgtcgtccctataaaaacggtaacaatgaaagctgt
      S  Y  D  C  R  P  Y  K  N  G  N  N  E  S  C
1846 tcgattaccgtcactcaagctggaacttaccatgtgatgttacgt
      S  I  T  V  T  Q  A  G  T  Y  H  V  M  L  R
1891 ggttatgctaattactcgggcgttcaactgagcgcaagctactag
      G  Y  A  N  Y  S  G  V  Q  L  S  A  S  Y  *

      cgagtaaatcagctcagtgagttttgactcacacttaagcccact

      aaggtaacaacttagtgggcttttttagtggtttattcagaagcca

      atcccgcgaaatggcattaatcacttgatcaataacttggagact

      tagcgaggataaccgttggttttgcttcgatttgcagacgctcaa

      tcactgcggcgagttttgcaatgctgaggcgaattggggt.....
```

Fig. 3-19: Nucleotide sequence and deduced amino acid sequence encoding the subtilisin-like serine protease gene from *Shewanella arctica*. The stop codon (TAG) is marked by an asterisk. The first 38 amino acids of the enzyme representing the signal sequence are indicated in bold letters. The three conserved amino acids Asp-His-Ser of the catalytic domain and the Asn residue of the oxyanion hole are indicated in bold underlined letters. The subtilisin-N conserved domain found between Asp-55 and Thr-135 residues is underlined.

The comparison of the amino acid sequence of the ORF2 protease with other proteases available in GenBank showed 65 % amino acid identity to the extracellular alkaline subtilisin-like serine protease from *Pseudoalteromonas* sp., 57 % identity to the subtilisin-like serine protease from *Alteromonas* sp. and 44 % identity with alkaline subtilisin-like serine protease from *Vibrio cholerae*. On the other hand, the ORF2 protease exhibit 36 % amino acid identity to a serine protease from the fungus *Arthrobotrys oligospora*. The identity and similarity percentage of the new subtilisin-like serine protease amino acid sequence from *Shewanella artica* with the most related serine proteases from 8 different bacterial strains and 4 fungal strains are shown in Table 3-10.

The multiple alignment of the amino acid sequences from the new subtilisin-like serine protease and the most related subtilisin-like serine proteases from *Pseudoalteromonas* sp. (BAB61726.1), *Vibrio cholerae* (NP229814.1), *Alteromonas* sp.(BAA18912.2), *Stenotrophomonas maltophilia* (AAP13815.1), *Xanthomonas campestris* (ATCC 33913) (NP_636242.1) and *Xanthomonas axonopodis* (NP_641280.1) is shown in Fig. 3-20. The protein has the three amino acid residues (Asp (D)-180, His (H)-213 and Ser (S)-364) that form the catalytic triad. This has been observed for the subtilisin-like serine proteases. High degree of homology near these residues at the amino acid level was observed in comparison with other subtilisin-like serine protease (Fig.3-21A). The highly conserved oxyanion region (Asn-299 residue of the oxyanion hole) which is also found in all subtilisin-like serine protease is also present (Fig.3-21B). At the N-terminal region a subtilisin-N conserved domain was found (GenBank database) between Asp-55 and Thr-135 residues, showing similarity to the N-terminal subtilisin domains of the subtilisin-like serine proteases from *Alteromonas* sp.(BAA18912.2) (Asp-36.....Thr-116), *Vibrio cholerae* (NP229814.1) (Asn-51.....Asn-131), *Vibrio alginolyticus* (P16588) (Asp-56.....Asp135), *Deinococcus radiodurans* (AAF11873) (Gly-88.....Ser-175), *Moraxella* sp. (AAD30204) (Gly-30.....Ser.109) and *Bacillus* sp. (1SPBP) (Lys-9.....Thr-78).

Construction of phylogenetic tree based on amino acids sequences from *Shewanella arctica* sp. nov. subtilisin-like serine protease and the most related serine proteases from the bacterial strains: *Pseudoalteromonas* sp., *Vibrio cholerae, Alteromonas* sp., *Stenotrophomonas maltophilla, Xanthomonas campestris, Xanthomonas axonopodis, Thermus* sp. and *Methylococcus capsulatus,* and from the fungal strains: *Arthrobotrys oligospora, Candida boidinii, Pyrenopeziza brassicae* and *Paracoccidioides brasiliensis* is shown in Fig.3-22.

```
Shewanella arctica              1  MSRSPWRHKKKMEFNMHKKHLIAVAVATGLAY---FPVNANEYQATMVSV
Pseudoalteromonas sp.           1  --------------MEHKHKIALAVLAALST---LPMQSQAAQLLPVNS
Vibrio cholerae                 1  --------MFKKFLSLCIVSTFSVAATSALAQPNQLVGKSSPQQLAPLMK
Alteromonas sp.                 1  --------MTINKNFKRAALSVAVSALFTATAVTAAPAQSIGQDAALAAT
Stenotrophomonas maltophilia    1  --------MSQVTQPRVRRVWVVLGASVLSSLLLATPALAGDVHSAGLQS
Xanthomonas campestris          1  --------MSTASLRKRTGSLTILGASALTSLLLAMPAFAGEVYLDGLAT
Xanthomonas axonopodis          1  --------MSNASFRPRHRALCILGVSALTSLLLVTPAFAGDVYLGGLAT


Shewanella arctica             48  PQ-SK----------A-IKDTYIVVFN------TPSVLNLSNNNTIAEFA
Pseudoalteromonas sp.          33  ---NK----------A-IKDTYIVVFN------TPSILNLQDTSALANFS
Vibrio cholerae                43  AASGK----------G-IKNQYIVVLK------QPTTIMSNDLQAFQQFT
Alteromonas sp.                43  AQKLSSQSTLGTQFIIKYKDASSQMMGLSAADLAPAKMQQRAESFVQDFK
Stenotrophomonas maltophilia   43  APTHQ-------RFIVKYRDGSAPVA-------NTTALASSLKSAAAGLA
Xanthomonas campestris         43  AQTHQ-------KFIVTYKDGSTALA-------SPSALTTSLRTAARAVP
Xanthomonas axonopodis         43  AQTHQ-------KFIVTYKDGSSALS-------SPSVLTTSLRTAARALP


Shewanella arctica             80  VQQAESLVNQYDVRVMKN---FGNVLNGVLINASAQQVKALLKDPNVKYV
Pseudoalteromonas sp.          63  TQQANILANDYNISVVKN---FGNALNGVLIKANAKQIAELQDDPKIKYI
Vibrio cholerae                76  QRSVNALANKHALEIKNV---FDSALSGFSAELTAEQLQALRADPNVDYI
Alteromonas sp.                93  SSK-ASARAQYVRPMALSNHHVMRADKKLSAKEAQAFMNEVVASGNVEYI
Stenotrophomonas maltophilia   79  SSQGRALGLQEVRKLAVG-PTLVRTDRPLDQAESELLMRKLAADPNVEYV
Xanthomonas campestris         79  AKAGKALGLNSVRRLALG-PELVRADRALDRAEAETLMRQLAADPNVQSV
Xanthomonas axonopodis         79  AKAGKSLGLNSVRRLAVG-PELVQADRALDRAEAETLMRQLAADPNVQSV


Shewanella arctica            127  EQDQVMSVTPMMEANA--DQPSPTWGIDRIDQRNLPLDNNYHTDYDGSGV
Pseudoalteromonas sp.         110  EQDQMMSISPIMSASS--DQGSPTWGLDRIDQRDLPFDNNYHYDYDGTGV
Vibrio cholerae               123  EQNQIITVNPIISASANAAQDNVTWGIDRIDQRDLPLNRSYNYNYDGSGV
Alteromonas sp.               142  EIDQML--KPFATPND--PRYNDQWHYYEAAAG-INAPAAWDKAT-GQGV
Stenotrophomonas maltophilia  128  EVDQIM--RATLTPND--TRLSEQWGFGTSNAS-INVRPAWDKAT-GTGV
Xanthomonas campestris        128  EVDQIL--HATLTPND--TRLSEQWAFGTTNAG-LNIRPAWDKAT-GSGT
Xanthomonas axonopodis        128  EVDQML--YPTLTPND--SRLSEQWAFGTTNAG-LNIRPAWDKAT-GANV


Shewanella arctica            175  TAFVIDTGVLNTHNEFGGRASSGYDFIDN----------DYDTTD-----
Pseudoalteromonas sp.         158  TAYIIDTGVLNSHNEFGGRATSGYDFIDN----------DSDTTD-----
Vibrio cholerae               173  TAYVIDTGIAFNHPEFGGRAKSGYDFIDN----------DNDASD-----
Alteromonas sp.               186  VVAVLDTGYR-PHLDLDANILPGYDMISNTFVANDGGARDNDARDPGDAV
Stenotrophomonas maltophilia  172  VVAVIDTGIT-NHPDLNANILPGYDFISDAAMARDGGGRDNNPNDEGDWY
Xanthomonas campestris        172  VVAVIDTGIT-SHADLNANILAGYDFISDATTARDGNGRDSNAADEGDWY
Xanthomonas axonopodis        172  VVAVIDTGIV-SHPDLDANILPGYDFISDATAARDGNGRDNNPADEGDWN
                                         *


Shewanella arctica            210  ----CNG------------------HGTHVAGTIGGST-----Y-GVAKNVN
Pseudoalteromonas sp.         193  ----CNG------------------HGTHVAGTIGGAS-----Y-GVAKNVN
Vibrio cholerae               208  ----CQG------------------HGTHVAGTIGGAQ-----Y-GVAKNVN
Alteromonas sp.               235  TRGECGTDSSGQPVPRADQDSSWHGTHVAGTVAAVTNNGEVGVGVAYDAK
Stenotrophomonas maltophilia  221  GANECGSGIPA-------SNSSWHGTHVAGTVAAVTNNSTGVAGTAFNAK
Xanthomonas campestris        221  AANECGAGIPA-------ASSSWHGTHVAGTVAAVTNNTTGVAGTAYGAK
Xanthomonas axonopodis        221  STSGCAT----------SNSSWHGTHVAGTVAALTNNTTGVAGTAFNAK
                                                            *
```

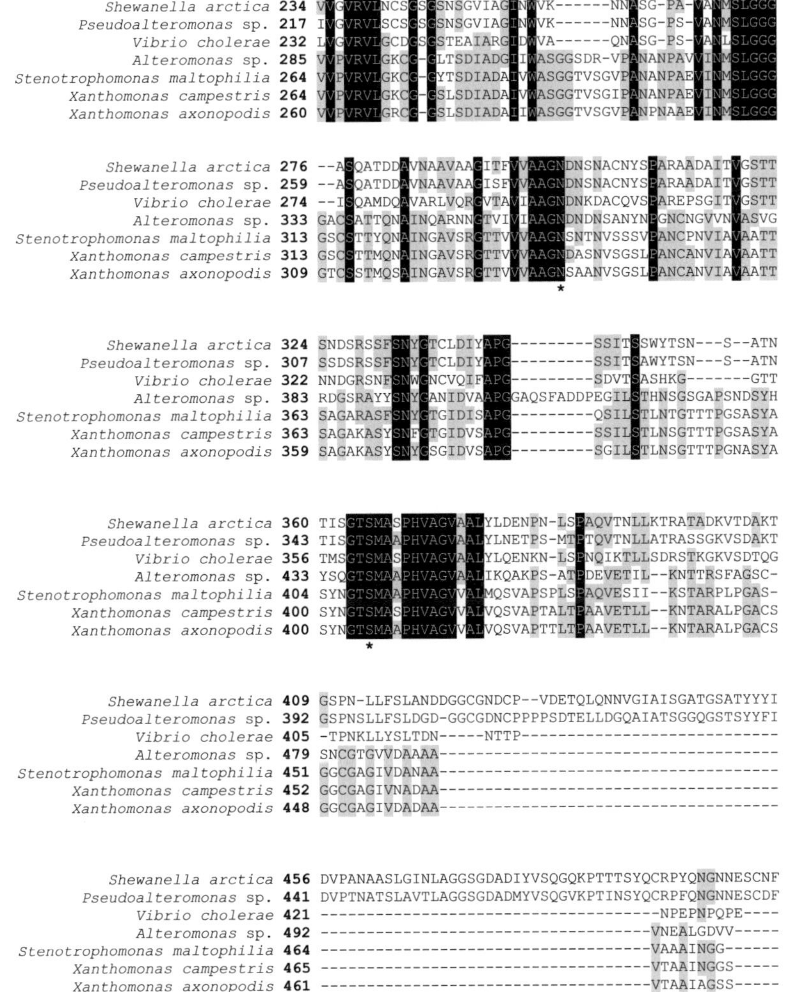

Shewanella arctica   234 VVGVRVLNCSGSGSNSGVIAGINWVK------NNASG-PA-VANMSLGGG
Pseudoalteromonas sp.   217 IVGVRVLSCSGSGSNSGVIAGINWVK------NNASG-PS-VANMSLGGG
Vibrio cholerae   232 LVGVRVLGCDGSGSTEAIARGIDWVA------QNASG-PS-VANLSLGGG
Alteromonas sp.   285 VVPVRVLGKCG-GLTSDIADGIIWASGGSDR-VPANANPAVVINMSLGGG
Stenotrophomonas maltophilia 264 VVPVRVLGKCG-GYTSDIADAIVWASGGTVSGVPANANPAEVINMSLGGG
Xanthomonas campestris   264 VVPVRVLGKCG-GSLSDIADAIVWASGGTVSGIPANANPAEVINMSLGGG
Xanthomonas axonopodis   260 VVPVRVIGRCG-GSLSDIADAIIWASGGTVSGVPANPNAAEVINMSLGGG

Shewanella arctica   276 --ASQATDDAVNAAVAAGITFVVAAGNDNSNACNYSPARAADAITVGSTT
Pseudoalteromonas sp.   259 --ASQATDDAVNAAVAAGISFVVAAGNDNSNACNYSPARAADAITVGSTT
Vibrio cholerae   274 --ISQAMDQAVARLVQRGVTAVIAAGNDNKDACQVSPAREPSGITVGSTT
Alteromonas sp.   333 GACSATTQNAINQARNNGTVIVIAAGNDNDNSANYNPGNCNGVVNVASVG
Stenotrophomonas maltophilia 313 GSCSTTYQNAINGAVSRGTTVVVAAGNSNTNVSSSVPANCPNVIAVAATT
Xanthomonas campestris   313 GSCSTTMQNAINGAVSRGTTVVVAAGNDASNVSGSLPANCANVIAVAATT
Xanthomonas axonopodis   309 GTCSSTMQSAINGAVSRGTTVVVAAGNSAANVSGSLPANCANVIAVAATT
                                                   *

Shewanella arctica   324 SNDSRSSFSNYGTCLDIYAPG---------SSITSSWYTSN---S--ATN
Pseudoalteromonas sp.   307 SSDSRSSFSNYGTCLDIYAPG---------SSITSAWYTSN---S--ATN
Vibrio cholerae   322 NNDGRSNFSNWGNCVQIFAPG---------SDVTSASHKG-------GTT
Alteromonas sp.   383 RDGSRAYYSNYGANIDVAAPGGAQSFADDPEGILSTHNSGSGAPSNDSYH
Stenotrophomonas maltophilia 363 SAGARASFSNYGTGIDISAPG---------QSILSTLNTGTTTPGSASYA
Xanthomonas campestris   363 SAGAKASYSNFGTGIDVSAPG---------SSILSTLNSGTTTPGSASYA
Xanthomonas axonopodis   359 SAGAKASYSNYGSGIDVSAPG---------SGILSTLNSGTTTPGNASYA

Shewanella arctica   360 TISGTSMASPHVAGVAALYLDENPN-LSPAQVTNLLKTRATADKVTDAKT
Pseudoalteromonas sp.   343 TISGTSMAAPHVAGVAALYLNETPS-MTPTQVTNLLATRASSGKVSDAKT
Vibrio cholerae   356 TMSGTSMASPHVAGVAALYLQENKN-LSPNQIKTLLSDRSTKGKVSDTQG
Alteromonas sp.   433 YSQGTSMAAPHVAGVAALIKQAKPS-ATPDEVETIL--KNTTRSFAGSC-
Stenotrophomonas maltophilia 404 SYNGTSMAAPHVAGVAALMQSVAPSPLSPAQVESII--KSTARPLPGAS-
Xanthomonas campestris   400 SYNGTSMASPHVAGVAALVQSVAPTALTPAAVETLL--KNTARALPGACS
Xanthomonas axonopodis   400 SYNGTSMAAPHVAGVAALVQSVAPTTLTPAAVETLL--KNTARALPGACS
                                      *

Shewanella arctica   409 GSPN-LLFSLANDDGGCGNDCP--VDETQLQNNVGIAISGATGSATYYYI
Pseudoalteromonas sp.   392 GSPNSLLFSLDGD-GGCGDNCPPPPSDTELLDGQAIATSGGQGSTSYYFI
Vibrio cholerae   405 -TPNKLLYSLTDN-----NTTP---------------------------
Alteromonas sp.   479 SNCGTGVVDAAAA------------------------------------
Stenotrophomonas maltophilia 451 GGCGAGIVDANAA------------------------------------
Xanthomonas campestris   452 GGCGAGIVNADAA------------------------------------
Xanthomonas axonopodis   448 GGCGAGIVDADAA------------------------------------

Shewanella arctica   456 DVPANAASLGINLAGGSGDADIYVSQGQKPTTTSYQCRPYQNGNNESCNF
Pseudoalteromonas sp.   441 DVPTNATSLAVTLAGGSGDADMYVSQGVKPTINSYQCRPFQNGNNESCDF
Vibrio cholerae   421 ----------------------------------NPEPNPQPE----
Alteromonas sp.   492 ----------------------------------VNEALGDVV-----
Stenotrophomonas maltophilia 464 ----------------------------------VAAAINGG------
Xanthomonas campestris   465 ----------------------------------VTAAINGGS-----
Xanthomonas axonopodis   461 ----------------------------------VTAAIAGSS-----

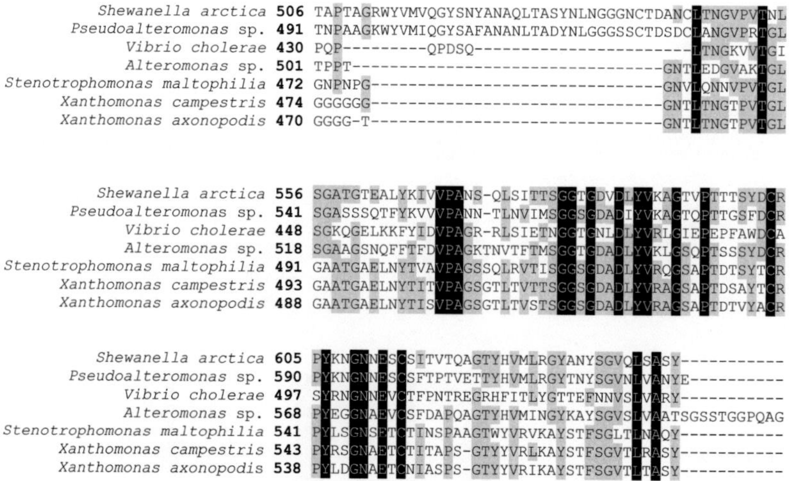

Fig. 3-20: Multiple alignments of amino acid sequences among the subtilisin-like serine protease from *Shewanella arctica* and the most related subtilisin-like serine proteases from *Pseudoalteromonas* sp. (BAB61726.1), *Vibrio cholerae* (NP229814.1), *Alteromonas* sp.(BAA18912.2), *Stenotrophomonas maltophilia* (AAP13815.1), *Xanthomonas campesTris* (NP_636242.1) and *Xanthomonas axonopodis* (NP_641280.1). The sequences were taken from the GenBank database. Completely identical conserved regions are marked by black boxes and conserved regions by grey boxes. The three conserved amino acids Asp-His-Ser of the catalytic domain and the Asn residue of the oxyanion hole are marked by an asterisk.

Table 3-10: Identity and similarity between *Shewanella arctica* subtilisin-like serine protease and those serine protease* from different bacteria and fungi

	Strain	GenBank accession no.	Identity (%)	Similarity (%)
Bacteria	*Pseudoalteromonas* sp.	BAB61726.1	65	77
	Alteromonas sp.	BAA18912.2	57	72
	Vibrio cholerae	NP229814.1	44	59
	Methylococcus capsulatus	YP_113369.1	41	55
	Thermus sp.	AAU14826.1	38	41
	Stenotrophomonas maltophilia	AAP13815.1	34	45
	Xanthomonas campestris	NP_636242.1	32	42
	Xanthomonas axonopodis	NP_641280.	32	41
Fungi	*Arthrobotrys oligospora*	CAA63841.1	36	49
	Candida boidinii	BAC75710.1	34	48
	Pyrenopeziza brassicae	CAC85639.1	32	46
	Paracoccidioides brasiliensis	AAP83193.1	32	44

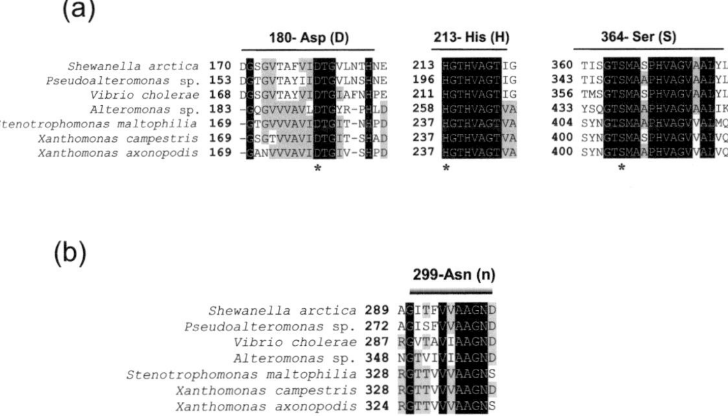

Fig. 3-21: Conserved regions around the three catalytic residues Asp(D)-His(H)-Ser(S) (a) and around the Asn(N) of the oxyanion hole (b). Completely identical conserved regions are marked by black boxes and conserved regions by gray boxes. The three catalytic residues and the Asn residue are marked by an asterisk. The numbers represents the positions of the amino acids residues starting from the N-terminus.

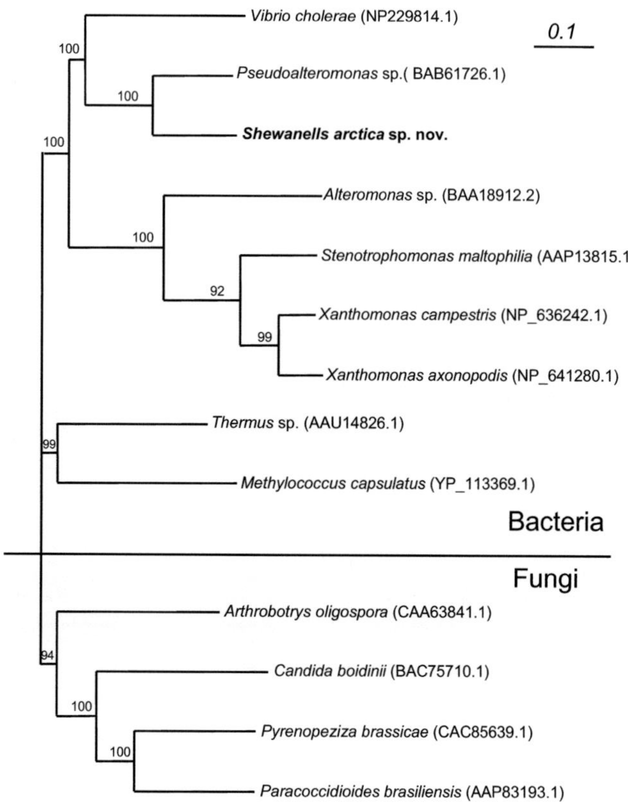

Fig. 3-22: Phylogenetic dendrogram based on comparison of amino acids sequences of the new subtilisin-like serine protease from *Shewanella arctica* and the related serine protease from different bacterial and fungal strains. Preformed by the neighbour-joining method using software from PHYLIP, version 3.57c; the PRODIST program with Kimura-2 factor was used to compute the pairwise evolutionary distances for the above aligned sequences , the topology of the phylogenetic tree was evaluated by performing a bootstrap (algorithm version 3.6 b) with 1000 bootstrapped trials. The tree was drawn using Tree View 32 software.

3.4.4 Purification of the recombinant protease after expression *E. coli* Tuner (DE3)pLacl.

Factors influencing the expression of recombinant protease from *E. coli* (DE3)pLacl were studied. Different concentrations of IPTG (0–5 mM) were used for induction, at various optical densities (OD_{600nm}= 0.6–1.2), at two different temperatures (30 °C and 37 °C) and at different incubation times (0–60 hour) after induction. Optimal expression conditions for the production of recombinant enzyme were determined by measuring the specific activity (U/mg) of the crude extract at the various conditions explained above. The highest specific activity (0.081 U/mg) was reached when *E. coli* (DE3)pLacl cells were grown on LB-carbenicillin-chloramphenicol medium, induced with 1mM IPTG when the cells optical density reached 0.9 and after 8 hours incubation at 30 °C (Fig. 3-23). Purification of the recombinant protease was carried out from the crude extract (as described in 2.5.3) from *E. coli* (DE3)pLacl cells grown at optimal expression conditions. The results of purification procedure is shown in Table 3-11. After Q-Sepharose chromatography the subtilisin-like serine protease was purified 6.93-fold, with specific activity of 0.52 U/mg and recovery yield of 8.67 %. Proteins from the purification step were separated using SDS-PAGE-8 % gel electrophoresis. The samples after Q-Sephaeose chromatography revealed one band with apparent molecular mass of 67.4 kDa. The zymogram staining showed that the same band is active on AZO-casein by forming clear zones (Fig. 3-24). The molecular mass of 67.4 kDa band matched the recombinant enzyme predicted from the recombinant protease amino acid sequence (Fig. 3-19). This result indicates that the protease is formed as a monomeric enzyme.

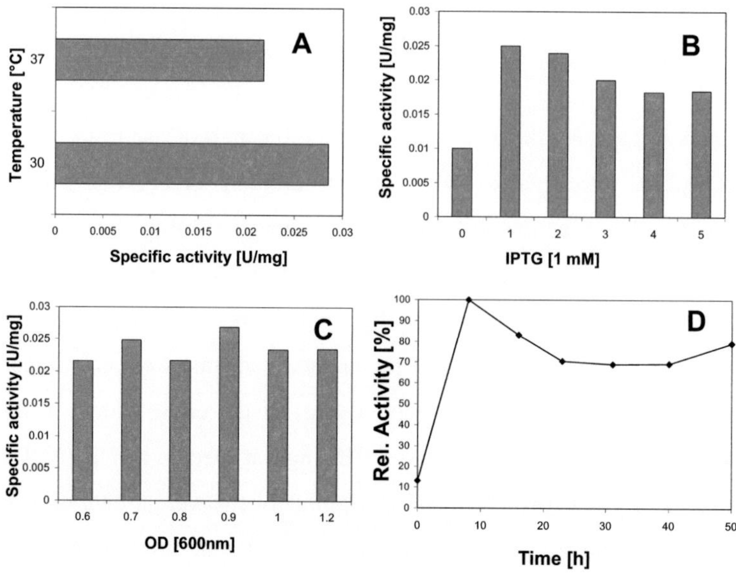

Fig. 3-23: Factors influencing the expression of recombinant protease from *E. coli* Tuner (DE3) pLacl. For determination of the optimum temperature (A) for the enzyme production, the cells were grown to OD 0.8, induced with 1 mM IPTG and incubated at 30 °C and 37 °C for 12 hours after induction. The enzyme production was tested by measuring the enzyme activity in cells crude extract. The influence of IPTG concentration (B) on the enzyme production was determined by growing the cell to OD 0.8 and induced with various IPTG concentrations (0 to 5 mM), at 30 °C for 12 hours. The enzyme production was tested by measuring the enzyme activity in the cells crude extract. The influence of induction (C) at various OD values (0.6-1.2), induced with 1 mM IPTG and incubated for 12 hours at 30 °C. For determination of the optimal incubation time (D) after induction, the cells were grown to OD 0.9, induced with 1mM IPTG and incubated at 30 °C for 50 hours. Samples were taken at various periods of time (0 to 50 h) and the enzyme activity was tested in the cells crude extract. LB-carbenicillin-chloramphenicol liquid medium was used as culture medium.

Table 3-11: Purification of the recombinant protease of *Shewanella arctica* after expression in *E. coli* Tuner (DE3) pLacI[$]

Purification step	Total protein (mg)	Total activity* (U)	Specific Activity (U/mg)	Recovery (%)	Purification (Fold)
Crude extract	26000	1950.9	0.075	100	1
Q-Sepharose	325	169.3	0.52	8.67	6.93

[$] After the aerobic growth of *E. coli* Tuner (DE3) pLacI at 30 °C, a 16 L culture was centrifuged (38 g of wet weight cells), cells were resuspended in Tris-NaCl-Triton X-100 buffer (pH 7), disrupted with French press, centrifuged and the supernatant was used for purification.

* One unit of protease is defined as the amount of enzyme that releases 1 μmol of aromatic amino acids per min under assay conditions specified. Tyrosine was used as a standard.

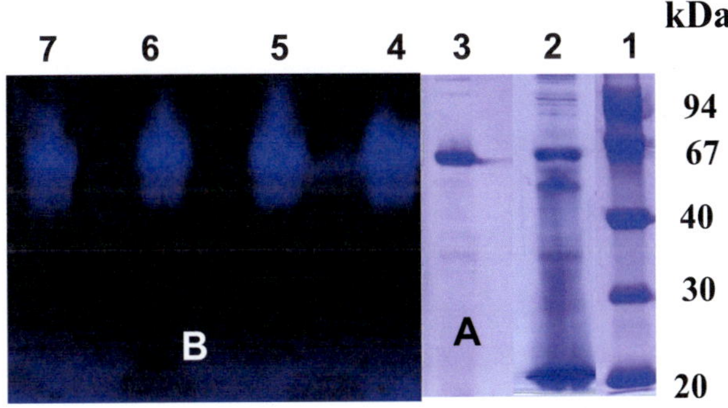

Fig. 3-24: Gel electrophoretic analysis (8 %-SDS-PAGE) and zymogram of the (67.4 kDa) recombinant protease. (A) Purification of recombinant protease. Lane 1, marker proteins; lane 2, crude extract (20 μg).; lane 3, Q-Sepharose pool (20 μg). (B) zymogram of the recombinant protease. Lane 4, 5, 6, and 7, Q-sepharose pool (20, 15, 10, and 5 μg respectively).

3.4.5 Biochemical properties and kinetic studies

The recombinant protease was active between 0-80 °C. The temperature optimum was 60 °C and a rapid decrease in the enzyme activity was observed above 60 °C (Fig. 3-25A). The recombinant protease showed activity over a broad pH range (pH 6–10), with an optimum at pH 8. A decrease in activity was observed above pH 8 (Fig. 3-25B). Temperature stability of the recombinant protease was examined by measuring the enzymatic activity after incubating the enzyme at low temperatures (R.T., 4, -20, and -80 °C) (Fig. 3-26B), and at high temperatures (40, 60, 70, 80, and 90 °C) (Fig. 3-26A). 50 % of the initial enzyme activity was detectable at -80 and -20 °C after 8 days of incubation. This loss of activity was detectable after 5 days of incubation at room temperature or 4 °C. The recombinant enzyme should stored at -80 °C and -20 °C rather than at 4 °C or room temperature. At 40 °C the enzyme was fully stable for 45 hours and exhibited a half-life of 1.5 hours at 70 °C. A complete lost of activity was observed after 30 min at 80 and 90 °C.

The apparent kinetic parameters according to Michaelis-Menten with different concentrations of casein (0.1 to 0.7 % (w/v)) were determined at pH 8 at 60 °C. The K_m was determined to be 0.175 % (w/v) and V_{max} 2500 U/l as shown in Fig. 3-27.

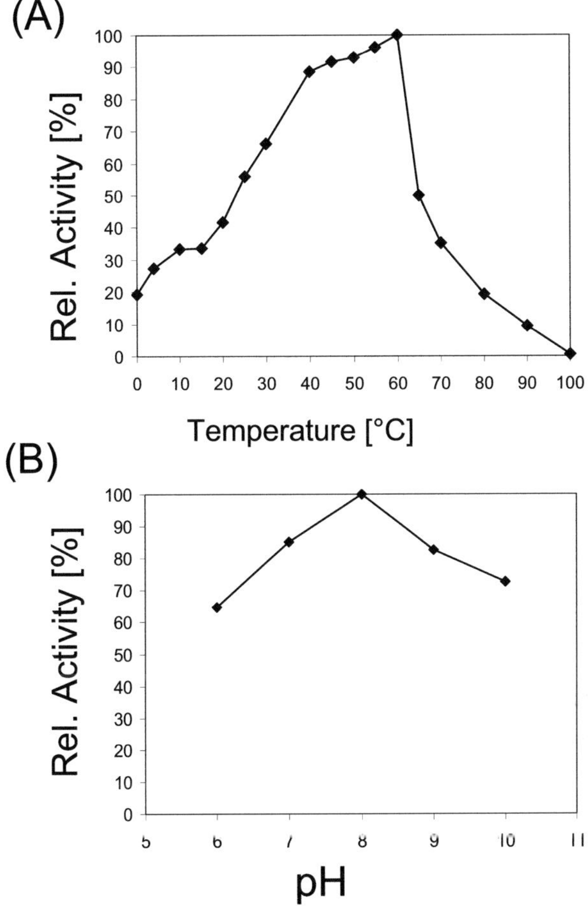

Fig. 3-25: Influence of temperature (A) and pH (B) on the activity of the recombinant protease from *Shewanella arctica*. For determination of temperature optimum, the purified enzyme (0.52 U/mg) was incubated at various temperatures (0 to 100 °C) for 10 min at pH 8. For determination of pH optimum, the purified enzyme was incubated in universal buffer at various pH values (pH 6 to 10) at 60 °C for 10 min.

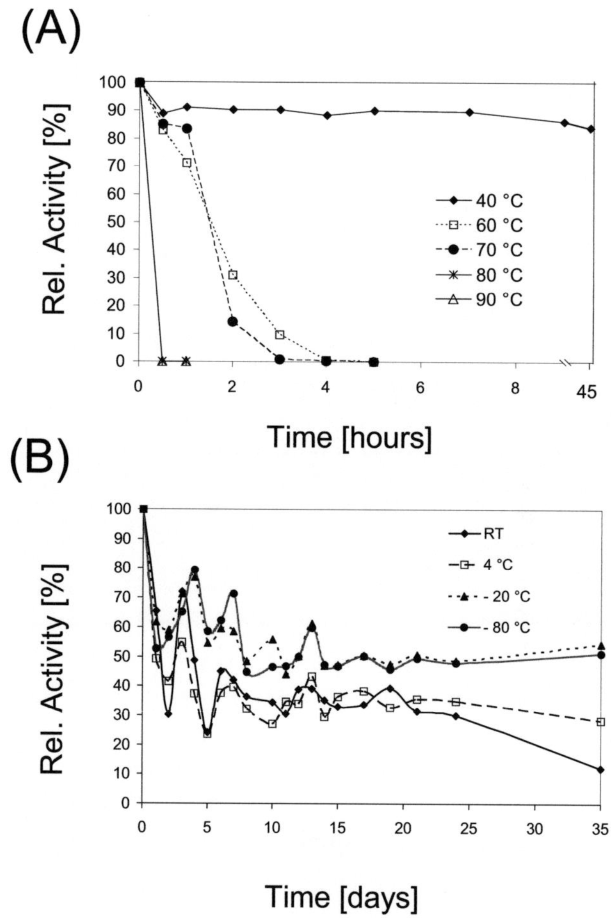

Fig. 3-26: Thermostability of the recombinant protease from *Shewanella arctica*. The enzyme (0.52 U/mg) was incubated at (A) 40 °C (♦), 60 °C (□), 70 °C (●) 80 °C (✕), and 90 °C (△). (B) room temperature (♦), 4 °C (□), -20 °C (▲), and -80 °C (●). Samples were withdrawn and tested for protease activity.

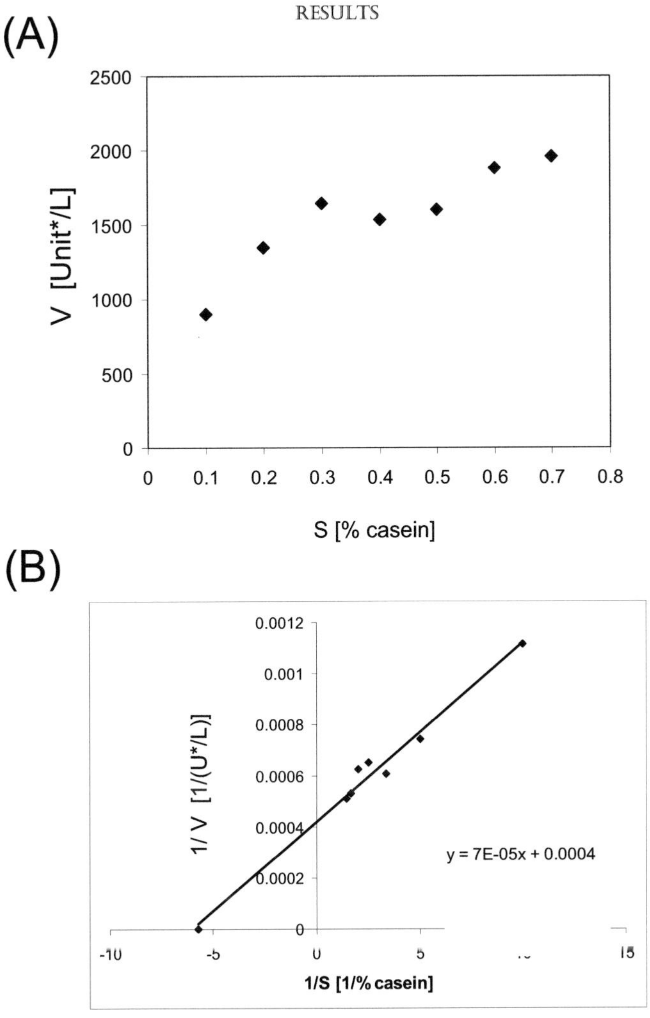

Fig. 3-27: Michaelis-Menten-kinetics of recombinant protease. (A) Effect of substrate concentration on the velocity of enzyme catalyzed reaction. (B) A double-reciprocal or Linewearer-Burk plot. In order to calculate the k_m and V_{max}, the enzyme activity was tested after (Kunitz, 1947) Kunitz (1947) at 60 °C, pH 8 and for 10 min with various concentrations of casein (0.1 to 0.7 % [w/v]). For each reaction 10 µl (0.53 U/mg) of pure enzyme solution was used. * One unit of protease is defined as the amount of enzyme that releases 1 µmol of aromatic amino acids per min under assay conditions specified. Tyrosine was used as a standard.

3.4.6 Effects of metal ions, reagents and inhibitors on the recombinant protease activity

All metal ions tested have no effect on the activity of the recombinant enzyme (Fig. 3-28A). Fe^{+3}, Co^{+2} and Wo_4^{-2} however, slightly increased the activity. High concentration of Ni^{+2} and Cu^{+2} caused a minimal decrease in the enzyme activity. All reagents tested did not significantly affect the enzyme activity (Fig. 3-28B). Only high concentrations (6 M) of urea and acetonitril significantly decreased the enzyme activity (< 40 % of initial activity). Bestatin (metallo protease inhibitor), leupeptin (serine and cysteine protease inhibitor), and pefabloc SC (serine protease inhibitor) did not inhibit the enzyme activity. Slight increase of enzyme activity was measured by phosphoramidon (thermolysin and neutral endopetidase inhibitor), aprotinin (serine protease inhibitor), and EDTA-Na_2 (metallo protease inhibitor). While, 37 % and 30 % loss of intial enzyme activity was measured by E-64 (cysteine and papain protease inhibitor) and pepstalin (aspartate protease inhibitor) respectively. Complete inhibition was observed by PMSF (phenylmethylsulfonyl flouride) (serine protease inhibitor), chymostatin (chymotrypsin-like serine and cysteine protease inhibitor), and antipain-dihydrochloride (serine and cysteine protease inhibitor). The effects of protease inhibitors on the new subtilisin-like serine protease are shown in Fig. 3-29 and Table 3-12.

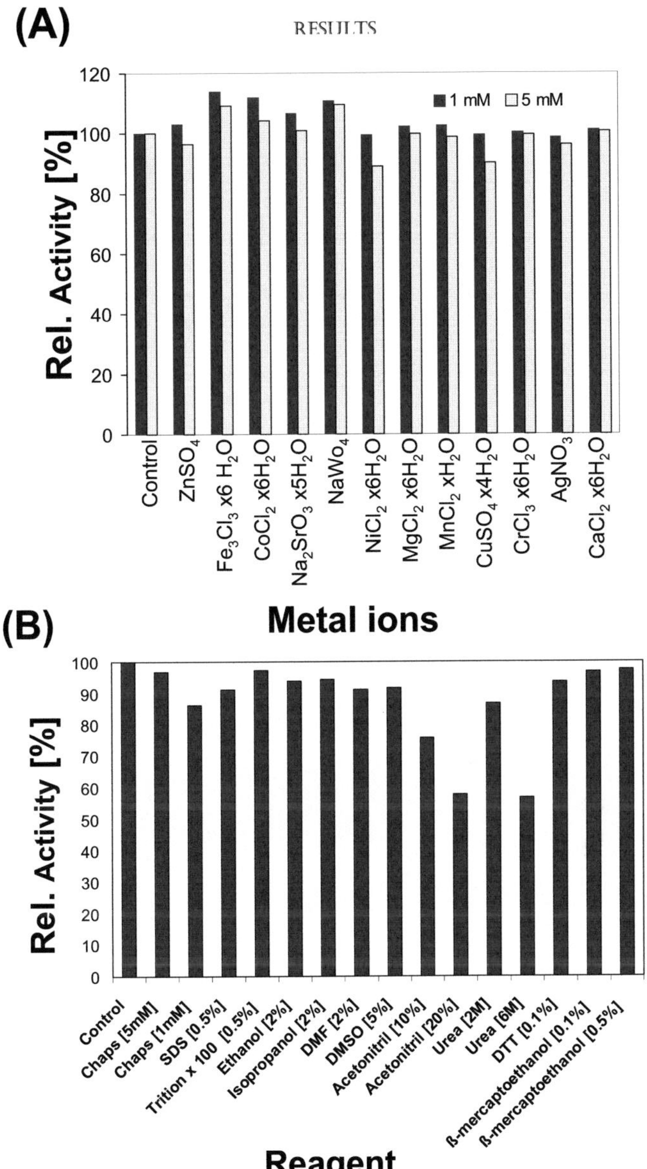

Fig. 3-28: Influence of different metal ions (A) and reagents (B) on recombinant protease activity. The effects of various reagents on the enzyme activity were examined after incubation of the recombinant protease (0.52 U/mg) with metal ions (1mM and 5 mM final concentration) and reagents (at various concentrations) at room temperature for 60 min. Samples were withdrawn and tested for protease activity at 60 °C for 10 min and pH 8. The enzyme activity without metal ions and reagents was considered as 100 % (control).

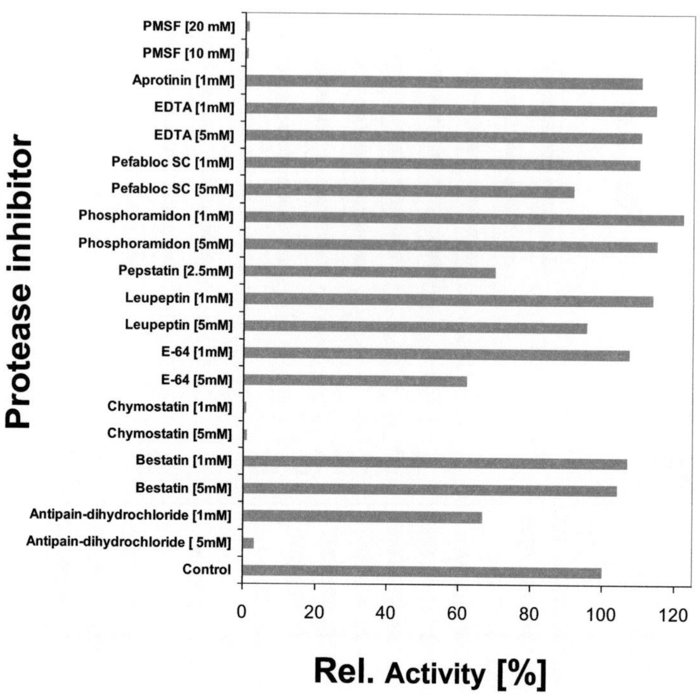

Fig. 3-29: Influence of different protease inhibitors on protease activity. The effects of various protease inhibitors on the enzyme activity were examined after incubation of the recombinant (0.52 U/mg) enzyme with inhibitors (at various concentrations) at room temperature for 60 min. Samples were withdrawn and tested for protease activity at 60 °C for 10 min and pH 8. The enzyme activity without inhibitors was considered as 100 % (control).

Table 3-12: Effects of protease inhibitors on the protease activity from *Shewanella arctica* sp nov.

Inhibitor[a]	Specificity of inhibitors	Concentration (mM)	Rel. Activity (%)
Control			100
Antipain-dihydrochloride	Reversible inhibitor of serine	5	3
	and cysteine proteases	1	66
Bestatin	Metalloprotease inhibitor with	5	103
	multi-pharmacological functions	1	106
Chymostatin	Reversible inhibitor of chymotrypsin-like	5	0.9
	serine and some cysteine proteases	1	0.6
E-64	Irreversible inhibitor of papain and	5	62
	other cysteine proteases	1	107
Leupeptin	Reversible competitive inhibitor of	5	95
	serine and cysteine protease	1	113
Pepstatin	Reversible inhibitor of aspartic proteases	2.5	69
Phosphoramidon	Specific inhibitor of thermolysin, endopeptidase	5	114
	and metallo-endoproteinases.	1	112
	(Does not inhibit serine, cysteine and aspartic proteases		
Pefabloc SC[b]	Irreversible inhibitor ofthrombin and other	5	91
	serine proteases	1	109
EDTA[c]	Reversible inhibitor of metalloproteases	5	110
		1	114
Aprotinin	Inhibits numerous serine proteases	1	110
PMSF[d]	Irreversibly inhibits serine proteases	10	0.7
		20	0.6

[a]=Roche diagnostics, Germany, [b]=4-(2.aminoethyl)-benzylsulfouride, [c]=ethylenediamine tetraacetate, [d]= phenylmethylsulfonyl fluoride.

4 DISCUSSION

4.1 *Ilyobacter psychrophilus* sp. nov., a novel anaerobic, psychrophilic bacterium isolated from Spitsbergen

The genus *Ilyobacter* consists so far of three species, *I. insuetus, I. tartaricus*, and *I. polytropus* sharing similar physiological and morphological characteristics and isolated from various origins (Brune *et al.*, 2002; Schink, 1984; Stieb & Schink, 1984). This genus was created to house mesophilic strictly anaerobic, gram negative, non motile, and non spore forming bacteria. *I. insuetus* was isolated with quinic acid as carbon source by Brune *et al*, (2002), *I. tartaricus* was isolated with L-, D-, and m- tartrate as carbon source by Schink (1984), and *I. polytropus* was isolated with 3-hydroxybutyrate as carbon source by Stieb & Schink (1984).

The strain FQ50 showed similarity in morphological characteristics to the strains of the *Ilyobacter-Propionigenium* group. The bacteria was: short rods, 0.5–2.5 μm wide, 0.2–2.5 μm long, gram negative, non motile, non spore forming, strictly anaerobic and occurred singly or in pairs (Janssen & Liesack, 1995; Schink, 1984; Schink & Pfenning, 1982; Stieb & Schink, 1984; Watson *et al.*, 2000) However, *Ilyobacter insuetus* cells were identified as coccoid in shape (Brune *et al.*, 2002).

The optimal pH and NaCl concentration (% (w/v)) of the strain FQ50 (pH 7 and 2 % NaCl (w/v)) were similar to those of *Ilyobacter-Propionigenium* group and optimal pH value range was 6–8 and NaCl % (w/v) was 0.5 – 2 % (w/v) (Brune *et al.*, 2002; Janssen & Liesack, 1995; Schink, 1984; Schink & Pfenning, 1982; Stieb & Schink, 1984; Watson *et al.*, 2000). On the other hand, the optimal temperature of the strain (FQ50) is not similar to those reported for *Ilyobacter-Propionigenium* group. All members of this group have an optimal temperature in the mesosphilic range (28–37 °C) (Table 3-1). The optimal growth temperature of the strain FQ50 was lower (10 °C). This psychrophilic behavior differentiates the new strain FQ50, from the strains of the *Ilyobacter-Propionigenium* group.

Phylogenetic treeing placed the strain FQ50 (according to its 16S rDNA composition) among *Ilyobacter-Propionigenium* group (Fig. 3-2) belonging to *Fusobateriaceae* family which contain seven genus; *Fusobacterium, Ilyobacter, Leptotrichia, Propionigenium, Sebaldella, Sneathia* and *Streptobacillus* (NCBI database). The strain FQ50 *Ilyobacter polytropus* and *Propionigenium maris* shared 92 % identity based on 16S rDNA sequence. The FQ50 strain showed a stable relative branching order within *Ilyobacter-Propionigenium* group and was branched in next to *Propionigenium maris* and *Ilyobacter polytropus* (Fig. 3-2). The range of overall 16S rDNA sequence identities of 92 % shared includes the threshold value for species separation (Stackebrandt & Goebel, 1994) and indicates that the new isolated strain FQ50 belongs to the *Ilyobacter-Propionigenium* group. The DNA base composition is 28.7 mol % which is quite similar to the values reported for *Ilyobacter tartaricus* (33.1 mol %), *Ilyobacter insuetus* (35.7 mol %), *Ilyobacter polytropus* (32.2 mol %), *Propionigenium maris* (40 mol %) and

Propionigenium modestum (33.9 mol %) (Brune *et al.*, 2002; Janssen & Liesack, 1995; Schink, 1984; Schink & Pfenning, 1982; Stieb & Schink, 1984; Watson *et al.*, 2000)

Sulfate, sulfur, thiosulfate and nitrate were not utilized as electron acceptors neither from the strain FQ50, nor from the members of *Ilyobacter-Propionigenium* group (Brune *et al.*, 2002; Janssen & Liesack, 1995; Schink, 1984; Schink & Pfenning, 1982; Stieb & Schink, 1984; Watson *et al.*, 2000).

The strain FQ50 grew on: pyruvate, starch, glucose, yeast extract, citrate, fructose, maltose, oxaloacetate, glutamate, cysteine, glycerol, xylose or 3 hydroxybbutyrate but not, on succinate. *Ilyobacter tartaricus, Ilyobacter insuetus* and *Ilyobacter polytropus* grew also on a large variety of substrates (Table 3-1) but not on succinate (Brune *et al.*, 2002; Schink, 1984; Stieb & Schink, 1984). *Propionigenium maris* and *Propionigenium modestum* showed ability to grow on different substrates besides their ability to grow on succinate (Table 3-1) (Janssen & Liesack, 1995; Schink & Pfenning, 1982; Watson *et al.*, 2000). This ability of *Propionigenium* species to grow on succinate differentiates them from *Ilyobacter* species (Brune *et al.*, 2002).

The strain FQ50 fermented glucose, starch and pyruvate forming butyrate, acetate and propionate. *Ilyobacter polytropus* ferments pyruvate, citrate and oxaloacetate, forming acetate and formate; glucose and fructose, forming acetate, formate and ethanol; malate and fumarate, forming acetate, formate and propionate (Stieb & Schink, 1984). *Ilyobacter insuetus* fermentation of quinic acid and shikimic acid resulted in the formation of acetate and butyrate (Brune *et al.*, 2002).

Ilyobacter tartaricus fermented L-tartrate, citrate, pyruvate, oxaloacetate, glucose, fructose and glycerol, forming acetate, formate and ethanol (Schink, 1984). *Propionigenium maris* ferments succinate, fumarate, pyruvate, citrate, fructose and other substrates, forming propionate, formate, butyrate, lactate and ethanol (Janssen & Liesack, 1995). And finally *Propionigenium modestum* fermented succinate, fumarate, aspartate, malate, pyruvate and oxaloacetate forming propionate and acetate (Schink & Pfenning, 1982).

We conclude that, *Ilyobacter* species and the strain FQ50 were not able to grow and metabolize succinate. However, the new strain FQ50 and *Ilyobacter polytropus* were able to produce propionate as a fermentation product by fermenting different substrates but not succinate (Stieb & Schink, 1984). On the other hand, both *Propionigenium maris* and *Propionigenium modestum* showed ability to grow and ferment succinate to propionate (Janssen & Liesack, 1995; Schink & Pfenning, 1982). This ability to grow by decarboxylation of succinate to propionate characterizes the genus *Propionigenium* (Janssen & Liesack, 1995; Schink, 1984). The new isolated strain FQ50 shares with the described *Ilyobacter* species the inability to utilize succinate by decarboxylation. This feature differentiates the strain from the described *Propionigenium* species. Based on the obtained results we propose to assign the newly isolated strain to the genus *Ilyobacter* as *Ilyobacter psychrophilus* sp. nov.

4.2 *Shewanella arctica* sp. nov., a novel psychrophilic bacterium isolated from Spitsbergen

The genus *Shewanella* comprises a ubiquitous group of gram negative, aerobic and facultatatively anaerobic γ- *Proteobacteria* (Garrity & Holt, 2001; Gauthier *et al.*, 1995; MacDonell & Colwell, 1985; Satomi *et al.*, 2003; Venkateswaran *et al.*, 1999). This genus is comprised of more than 25 species inhabiting a wide range of environments including spoiled food oil field wastes, redox interfaces in marine and freshwater, cold water and sediments of the deep sea, and mesophilic representatives all around the planet (Satomi *et al.*, 2003). During the last decade, the representatives of this genus have received a significant amount of attention due to their important roles in co-metabolic bioremediation of halogenated organic pollutants (Petrovskis *et al.*, 1994), destructive souring of crude petroleum (Semple & Westlake, 1987) and the dissimilatory reduction of magnesium and iron oxides (Myers & Nealson, 1988).

The strain 40–3 showed high similarity in morphological characterization to the strains belonging to the genus *Shewanella* identified as: straight or curved rod–shaped, 2-3 μm length, 0.4-0.7 wide, gram negative and non–spore forming (Bowman *et al.*, 1997; Bozal *et al.*, 2002; Brettar *et al.*, 2002; Ivanova *et al.*, 2004a; Ivanova *et al.*, 2004b; Ivanova *et al.*, 2003a; Ivanova *et al.*, 2001; Ivanova *et al.*, 2003b; Leonardo *et al.*, 1999; Makemson *et al.*, 1997; Nogi *et al.*, 1998; Reid & Gordon, 1999; Satomi *et al.*, 2003; Skerratt *et al.*, 2002; Venkateswaran *et al.*, 1998; Venkateswaran *et al.*, 1999; Yoon *et al.*, 2004).

The optimal growth temperature and pH of the strain 40–3 (10–15 °C and pH 7–8) were similar to the most related *Shewanella* species: *Shewanella putrefaciens* (25–35 °C and pH 7-8), *Shewanella frigidimarina* (20–22 °C), *Shewanella pacifica* (20–25 °C), *Shewanella gaetbuli* (30 °C and pH 7-8) and *Shewanella waksmanii* (20–22 °C and pH 7.5) (Bowman *et al.*, 1997; Ivanova *et al.*, 2004a; Ivanova *et al.*, 2003a; Venkateswaran *et al.*, 1999; Yoon *et al.*, 2004). However, *Shewanella baltica* grew at 4 °C (Ziemke *et al.*, 1998). The strain 40–3 grew in the presence of a wide range of NaCl concentration from 0 to 10 % (w/v) (optimum 8 to 9 % (w/v)) unlike the related *Shewanella* species: *S. putrefaciens, S. frigidimarina, S. pacifica, S. gaetbuli* and *S. waksmanii,* where no growth was observed above 6 % NaCl (w/v) (Bowman *et al.*, 1997; Ivanova *et al.*, 2004a; Ivanova *et al.*, 2003a; Venkateswaran *et al.*, 1999; Yoon *et al.*, 2004).

The strain 40–3 is able to utilize a large number of substrates: α-cyclo-dextrin, dextrin, tween 80, N-acetyl-D-glucosamine, α-D-glucose, maltose, sucrose, methylpyruvate, D,L-lactitate, succiniate , bromo succinic acid, inosine, esculin ferric citrate, L-arabinose, potassium gluconate, malic acid and trisodium citrate. *Shewanella putrefaciens* and the strain 40–3 shared the ability to utilize: maltose, sucrose, succiniate, D–galactose, fumarate and lactose (Venkateswaran *et al.*, 1999). *Shewanella baltica* and the strain 40–3 on the other hand, shared the ability to utilize maltose, sucrose, malic acid and citrate (Ziemke *et al.*, 1998). The strain 40–3 and the related strains *Shewanella putrefaciens, Shewanella baltica, Shewanella frigidimarina* and *Shewanella waksmanii* were able to produce H_2S (from sodium thiosulfate) and reduce nitrates to nitrites (Bowman *et al.*, 1997; Ivanova *et al.*, 2003a; Venkateswaran *et al.*, 1999; Ziemke *et al.*, 1998). And finally the strain

40-3 and all previous *Shewanella* species were unable to produce indole or acetoin (Bowman *et al.*, 1997; Ivanova *et al.*, 2003a; Venkateswaran *et al.*, 1999; Yoon *et al.*, 2004).

The strain 40–3 produced several enzymes: amylase, pullulanase, protease, ornithine decarboxylase, alkaline phospatase, esterase (C4), esterase/lipase (C8), leucine arylamidase, valine arylamidase, trypsin, α-chymotrpsin, naphthol-AS-BI-phosphohydrolase and N-acetyl-β–glucosaminidase. Amylase was also produced by *Shewanella pacifica* and *Shewanella gaetbuli* and protease was also produced by *Shewanella pacifica* (Ivanova *et al.*, 2004a; Yoon *et al.*, 2004).

The strain 40-3 has 17.89 % total straight chain saturated FAME, 14.89 % terminally branched saturated FAMEs and 17.73 % total monounsaturated FAMEs. 16:0 straight chain saturated FAME was the most abundant FAME found in the strain 40-3 (11.24 %) as well as, in the related strains: *Shewanella putrefaciens* (19.1 %), *Shewanella baltica* (4.3 %), *Shewanella frigidimarina* (11.8 %), *Shewanella gaetbuli* (8.4 %) and *Shewanella waksmanii* (6.2 %) (Bowman *et al.*, 1997; Ivanova *et al.*, 2004a; Ivanova *et al.*, 2003c; Venkateswaran *et al.*, 1999; Yoon *et al.*, 2004). 15:0-iso terminally branched saturated FAME was the most abundant FAME found in the strain 40-3 (9.4 %) and in the related strains: *Shewanella putrefaciens* (21.1%), *Shewanella baltica* (14.3 %), *Shewanella frigidimarina* (9 %), and *Shewanella waksmanii* (32.5 %) (Bowman *et al.*, 1997; Ivanova *et al.*, 2004a; Ivanova *et al.*, 2003a; Venkateswaran *et al.*, 1999). However, 16:1 ω7c was the most abundant FAME in all *Shewanella* species mentioned above, but not in the strain 40–3 where, 18:1 ω7c was the most abundant FAME (Table 3-4).

High similarity was shown between the G + C content of strain 40-3 (48 mol %) and related strains: *Shewanella putrefaciens* (47 mol %), *Shewanella baltica* (46 mol %), *Shewanella frigidimarina* (40-43 mol %), *Shewanella pacifica* (40 mol %), *Shewanella gaetbuli* (42 mol %) and *Shewanella waksmanii* (42 mol %) (Bowman *et al.*, 1997; Ivanova *et al.*, 2004a; Ivanova *et al.*, 2004b; Ivanova *et al.*, 2003a; Venkateswaran *et al.*, 1999; Yoon *et al.*, 2004). Phylogenetic treeing placed the strain 40-3 (according to its 16S rDNA composition) among the species of the genus *Shewanella* (about 33 species). The strain 40-3 shared (according to its 16S rDNA composition) 99 % identity with *Shewanella putrefaciens*, 98 % identity with *Shewanella massilia* and *Shewanella baltica*, 97 % identity with *Shewanella decolorationis* and *Shewanella frigidimarina*, 96 % identity with *Shewanella pacifica* and *Shewanella gaetbuli,* and finally 95 % identity with *Shewanella waksmanii* and *Shewanella marisfiavi*. The strain 40-3 showed a stable relative branching order with *Shewanella putrefaciens* and was placed in between *Shewanella putrefaciens* and *Shewanella massilia* (Fig. 3-4). DNA-DNA hybridization of the strain 40-3 and the most identical strain *Shewanella putrefaciens* showed only 50 % DNA–DNA similarity despite the fact that the 16S rDNA identity between both, the new strain and *Shewanella putrefaciens* was very high (99 %). This strongly indicates, that the newly isolated strain (40-3) is a new species within the genus *Shewanella*.

All data mentioned previously demonstrate clearly that the new isolate 40-3 is a new species within the genus *Shewanella*. This fact was also supported by the DNA-DNA hybridization, phylogenetic, G+C content, morphological and fatty acid analysis data. Based on the phylogenetic analysis by 16S rDNA and DNA-DNA hybridization homology, the new isolated strain 40-3 was found to be closely related to *Shewanella putrefaciens*, but represents a new species within the genus *Shewanella*. we propose to assign the newly isolate to the genus *Shewanella* as *Shewanella arctica* sp. nov..

4.3 Cloning, sequencing, purification and characterizeation of a pullulanase type I from *Shewanella arctica* sp. nov.

In recent years the number of reports on enzymes from psychrophilic bacteria has grown significantly. Of special interest is their high catalytic efficiency and capability to work at low temperature. They are therefore potential targets for novel industrial application (Leiros *et al.*, 2000). Because of their potential biotechnological application psychrophilic bacteria are now receiving increased attention (Aghajari *et al.*, 1998). Indeed, potential application can be found for cold-adapted enzymes in industrial processes (as additives in detergents), in food industries or in bioremediation (Aghajari *et al.*, 1998). The conversion of starch to glucose involves the debranching enzymes and glucoamylase.

A gene encoding type I pullulanase form the psychrophilic bacterium *Shewanella arctica* sp nov. was cloned and expressed as an active enzyme in *E. coli* by screening genomic library. The gene sequence (4.323 kb, 47.97 % GC and 1440 aa) showed high homology to those of type I pullulanase. The insert of 6-kb contains only one open reading frame (ORF) encoding a protein of 1440 amino acids length with molecular mass of 155.9 kDa. Downstream of the Shine-Dalgaron-like sequence (AGGAGA) the sequence started with ATG as translational initiation codon. Computer analysis suggests that the recombinant pullulanase possesses a signal sequence with 23 amino acids and 2.3 kDa

Analysis and alignment of the amino acid sequence carried out by comparing *Shewanella arctica* pullulanase and related enzymes revealed the presence of FNWGYDP motif that is common to all type I pullulanases which can only hydrolyze α-1,6 bonds in pullulan (Bertoldo & Antranikian, 2002a,b; Bertoldo *et al.*, 2004; Bertoldo *et al.*, 1999; Duffner *et al.*, 2000). Also the four conserved regions (regions I, II, III, and IV) which are typical of all amylolytic enzymes that belong to glycosyl hydrolase family 13 (the so-called α amylase family) were found (Table 3-7) (Bertoldo *et al.*, 2004; Bertoldo *et al.*, 1999). The four conserved regions are involved in the active site architecture within the catalytic domain of glycosyl hydrolase family 13 enzymes and encompass the catalytic residues (Nakajima *et al.*, 1986). The residues Asp-642, Glu-747 and Asp-856 of pullulanase form *Shewanella arctica* correspond to Asp-636, Glu-741 and Asp-841 of *Microbulbifer degradans* pullulanase (GenBank) and to Asp-608, Glu-714 and Asp-841 of *Klebsiella pneumoniae* pullulanase (Michaelis *et al.*, 1985). Other amino acids such as Trp-749 and Trp-860 residues of the *Shewanella arctica* pullulanase correspond to Trp-743 and Trp-847 residues of *Microbulbifer degradans* pullulanase (GenBank) and Trp-716 and Trp-847 residues for *Klebsiella pneumoniae* pullulanase (Michaelis *et al.*, 1985), in regions III and IV, respectively. They can be involved in substrate binding or catalysis (Bertoldo *et al.*, 2004). The Trp-749 residue for the *Shewanella arctica* pullulanase corresponds to Trp-497 in region III of isoamylase of *Pseudomonos amyloderamosa* (Amemura *et al.*, 1988; Katsuya *et al.*, 1998), which specifically hydrolyzes α- 1,6 glycosyl linkages in starch. In addition, His residues, which have been found to participate specifically in the hydrolysis of α-1,6 glycosyl linkages by *Klebsiella aerogenes* pullulanase (Michaelis *et al.*, 1985), are also present in *Shewanella arctica* pullulanase in region II and IV at position His-722 and His-854, respectively. The new *Shewanella arctica* pullulanase exhibits 53 % amino acid identity with *Microbulbifer degradans* pullulanase, 39 % with *Klebsiella aerogenes* pullulanase, and 40 % with *Chloroflexus aurantiacus* pullullanase (Table 3-6), which are all classified as pullulanase type I (Michaelis *et al.*, 1985). The phylogenentic analysis of *Shewanella arctica* pullulanase based on amino acid sequence placed the enzyme next to *Microbulbifer degradans* type I pullulanase (Fig. 3-7).

The activity of the pullulanase produced and purified from recombinant *E. coli* crude extract (2.7 U/mg) was similar to that from the native purified enzyme produced from *Shewanella arctica* cells crude extract (3 U/mg). The native pullulanase was purified to homogeneity, allowing some characterization of structural, physiochemical properties. The N-terminal sequence (CGGSDSSEPGKVLLTCDV) matched the predicted N-terminal sequence from the *Shewanella arctica* pullulanase gene. SDS-PAGE anylisis showed that the molecular mass of the native pullulanase is 155.9 kDa. Four purification steps were necessary to purify the recombinant pullulanase. SDS-PAGE and zymogram experiments from the second purification step (hydroxylapatite chromatography) showed that the recombinant enzyme is composed of two polypeptides of apparent molecular masses of 85 and 70 kDa. Interestingly, the sum of the molecular mass of both proteins matched the size of the active enzyme. Since SDS-PAGE and zymogram analysis of the native purified pullulanase show that the native (155.9 kDa) enzyme is a monomer, the two active recombinant proteins bands (85 and 70 kDa) indicate, that probably the recombinant enzyme has been hydrolyzed in *E. coli* in two polypeptides both active as pullulanase. The monomeric quaternary structure resembles that of the other debranching enzymes described so far (Albertson *et al.*, 1997; Bertoldo *et al.*, 2004; Bertoldo *et al.*, 1999; Bibel *et al.*, 1998; Messoud *et al.*, 2002; Michaelis *et al.*, 1985; Suzuki *et al.*, 1991), with the exception of *Fervidobacterium pennirorans* pullulanase which has a dimeric structure (Bertoldo *et al.*, 1999). The characterization of structural and physiochemical properties of the recombinant pullulanase has been carried out with the enzyme sample from the second purification step (hydroxylapatite chromatography). At this stage the enzyme was purified to 7-fold, with 11.3 % recovery and specific activity increased from 0.39 U/mg to 2.7 U/mg. The molecular mass of *Shewanella arctica* pullulanase (155.9 kDa) is very close to the values reported for *Microbulbifer degradans* pullulanase (155 kDa) (GenBank). Values from 65 to 140 kDa have been reported for the enzymes derived from *Klebsiella aerogenes*, *Chloroflexus aurantiacus*, *Vibrio vulnificus*, *Caldicellulosirupto saccharolyticus*, *Thermotoga maritime*, and *Bacillus acidopullulyticus* (Albertson *et al.*, 1997; Bibel *et al.*, 1998; Kuriki *et al.*, 1988; Michaelis *et al.*, 1985).

The optimal pH of the native and recombinant *Shewanella arctica* pullulanase activity (pH 7) is slightly higher than most pullulanase, where, pH optimum ranges between 5.5-6.0, such as *Bacillus stearothermophilus*, *Thermotoga caldophilus*, *Thermotoga maritime*, and *Bacillus thermoleovorans* (Altschul *et al.*, 1990; Bernfeld, 1955; Bibel *et al.*, 1998; Duffner *et al.*, 2000; Kim *et al.*, 1996; Messoud *et al.*, 2002). Both native and recombinant pullulanases are not optimally active in alkaline pH range (pH 8-10), as reported for the pullulanase from *Bacillus* sp. and *Anaerobranca gottschalkii* (Bertoldo *et al.*, 2004; Kim *et al.*, 1996). The optimum temperature (45 °C) of the native and recombinant *Shewanella arctica* pullulanase is different from all reported type I pullulanases, where most of them have optimal temperatures between 60 to 85 °C (Bertoldo *et al.*, 2004; Bertoldo *et al.*, 1999; Duffner *et al.*, 2000; Messoud *et al.*, 2002; Rudiger *et al.*, 1995). The recombinant type I pullulanase was stable at 40 °C for 3 h with a half-life of 44 min at 50 °C. The ability of this enzyme to be active at a wide range of temperature (4-60 °C) makes it an attractive candidate in industrial processes running at elevated temperatures.

Recombinant and native *Shewanella arctica* pullulanase are inhibited by the metal ions Mn^{+2}, Zn^{+2}, Cu^{+2} and Fe^{+2}. In contrast, Zn^{+2} was reported to inhibit *Anaerobranca gottschalkii* pullulanase, Zn^{+2}, Cu^{+2} and Fe^{+2} were reported to inhibit *Fervidobacterium pennavorus* pullulanase type I, *Pyrococcus woesei* and *Desulfurococcus mucosus* pullulanase type II. Complete inhibition of native and recombinant *Shewanella arctica* pullulanase was observed with EDTA, unlike the reported data, where EDTA was not inhibitory (Bertoldo *et al.*, 2004; Bertoldo *et al.*, 1999; Duffner *et al.*, 2000; Rudiger *et al.*, 1995). α-, β-, and γ- cyclodextrins and N- bromosuccinimide did not affect the activity of the native and recombinant enzymes. This is in contrast to most pullulanases whose activity is completely eliminated with N- bromosuccinimide (Bertoldo *et al.*, 2004; Bertoldo *et al.*, 1999; Duffner *et al.*, 2000). On the other hand, the enzyme stability with α-, β-, and γ- cyclodextrins was reported for pullulanase form *Desulfurococcus mucosus* (Duffner *et al.*, 2000). However, strong inhibition of pullulanase by α-, β-, and γ- cyclodextrins was reported for pullulanase from *Pyrococcus woesei, Fervidobacterium pennavorans,* and *Anaerobranca gottschalkii* (Bertoldo *et al.*, 2004; Bertoldo *et al.*, 1999; Duffner *et al.*, 2000; Rudiger *et al.*, 1995). SDS (sodium dodecysulfate), urea, DTT (dithiothreitol), guanidine-HCl and iodocetamide inhibited the recombinant and native enzyme activity.

The hydrolysis pattern after the recombinant pullulanase action on pulluan reveals the complete conversion of pullulan to maltotriose in end-acting fashion. The production of glucose after incubation of the hydrolysis products with α- glycosidase confirms, that the pullulanase hydrolyzed α-1,6 glycosyl linkages in pullulan producing maltotriose (possessing α-1,4 glycosyl linkages) (Fig. 3-15). Similar results were also reported for type I pullulanase from *Fervidobacterium pennavorans,* and *Bacillus thermoleovorans* (Bertoldo *et al.*, 2004; Bertoldo *et al.*, 1999; Duffner *et al.*, 2000; Messoud *et al.*, 2002; Rudiger *et al.*, 1995).

The ability of *Shewanella arctica* pullulanase to hydrolyze only α-1,6 glycosyl linkages in pullulan producing maltotriose confirmed by HPLC analysis, together with the multiple alignment and phylogenetic amino acids analysis classifies the new enzyme as a pullulanase type I. In conclusion, a novel type I pullulanase from the psychrophilic bacterium

Shewanella arctica sp. nov. was discovered (Patent registration number 102004046116.3). The enzyme was cloned and successfully expressed as an active enzyme in *E. coli*. Maximal pullulytic activity was found at 45 °C and pH 7. The enzyme was active between 4 to 60 °C.

4.4 Cloning, sequencing, purification and characterizeation of a subtilisin-like serine protease from *Shewanella arctica* sp. nov.

Serine proteases are of considerable interest, in view of their activity and stability at neutral and alkaline pH and low and high temperatures (0-95 °C). They have many applications in a number of industries. They have a nuleophilic serine residue located in their active site. Apart from this, these proteases are also distinguished by having essential aspartate and histidine which, along with the serine, forms the catalytic triad (Gupta *et al.*, 2002). Subtilisn-like serine proteases are generally of bacterial origin, although they were found in archaea and eukaryotes (Siezen & Lcunisscn, 1997). This family of proteases is specific for aromatic or hydrophobic residues (at position P1), such as tyrosine, phenylalanine and leucine. They are highly sensitive towards phenylmethylsulphonyl-fluoride (PMSF) and diisopropyl-fluorophosphate inhibitor (Gupta *et al.*, 2002). The serine proteases are most active around pH 8, with a molecular weight range of 15-115 kDa. Serine proteases are well represented in various bacterial strains such as: *Pseudoalteromonas* sp., *Shewanella* sp.(strain AC10), *Streplomyces albogriseolus*, *Mycobacterium tuberculosis*, *Xenorhabdus nemotophila*, *Thermoanaerobacter yonseiensis*, *Bacillus halmopalus*,

Fervidobacterium pennivorans and *Fervidobacterium isladicum* (Caldas *et al.*, 2002; Dave *et al.*, 2002; Gödde *et al.*, 2005; Jang *et al.*, 2002; Kluskens *et al.*, 2002; Kulakova *et al.*, 1999; Lee *et al.*, 2000; Saeki *et al.*, 2002; Suzuki *et al.*, 1997). Some proteases have been also detected or isolated from archaea such as: *Thermococcus kadaeaensis*, *Aeropyrum pernix* and *Natrialba magadii* (Catara *et al.*, 2003; Gimenez *et al.*, 2000; Kannan *et al.*, 2001), and some fungi such as *Pleurotus ostreatus* (Palmieri *et al.*, 2001).

A gene encoding a subtilisin-like serine protease was identified and sequenced from a gene library of the psychrophilic bacterium *Shewanella arctica* sp. nov.. The gene was amplified by PCR, sub-cloned and activity expressed in *E. coli* Tuner (DE3) pLacl. The gene sequence showed high homology to those of the subtilisin-like family, the largest class of serine proteases (Fig.1-7). In the sequenced region there are two open reading frame (ORF1 and ORF2) encoding two proteins. ORF2 was successfully sub-cloned and expressed as an active enzyme in *E. coli* Tuner (DE3) pLacl. This sub-cloned ORF2 encodes a protein of 644 amino acids length with a molecular mass of 67.4 kDa. Computer analysis suggests that the recombinant subtilisin-like protease possesses a signal peptide with 38 amino acids length and 4.4 kDa.

Analysis and alignment of the amino acid sequence of *Shewanella arctica* subtilisin-like gene and related subtilisins-like reveals the three amino acid residues Asp-180, His-213 and Ser-364 that are the catalytic triad characterizing this family (Carter & Wells, 1988). These amino acid residues have been found in all subtilisin-like proteases from the bacterial strains *Pseudoalteromonas* sp.(Asp-63, His-196 and Ser-348),

Vibrio cholerae (Asp-178, His-211and Ser-561), *Alteromonas* sp. (Asp-192, His-258 and Ser-438), *Stenotrophomonae maltophilia* (Asp-178, His-237 and Ser-409), *Xanthomonas campesTris* (Asp-178, His-237 and Ser-405), *Xanthomonas axonopodis* (Asp-178, His-237 and Ser-405), *Shewanella* sp. (strain AC10) (Asp-30, His-65 and Ser-369), *Streplomyces albogriseolus* (Asp-29, His-61 and Ser-238), *Mycobacterium tuberculosis* (Asp-90, His-61 and Ser-332) (Dave et al 2002), *Thermoanaerobacter yonseiensis* (Asp-29, His-64 and Ser-252), *Bacillus halmopalus* (Asp-245, His-283 and Ser-351), *Fervidobacterium pennivorans* (Asp-41, His-79 and Ser-260), and *Fervidobacterium isladicum* (Asp-177, His-215 and Ser-391) (Dave *et al.*, 2002; Gödde *et al.*, 2005; Jang *et al.*, 2002; Kluskens *et al.*, 2002; Kulakova *et al.*, 1999; Lee *et al.*, 2000; Saeki *et al.*, 2002; Suzuki *et al.*, 1997) (GenBank) the archaeon strains *Thermococcus kadaeaensis* (Asp-33, His-71 and Ser-242), *Aeropyrum pernix* (Asp-36, His-83 and Ser-321) (Catara *et al.*, 2003; Kannan *et al.*, 2001). In the catalytic reaction of subtilisin-like serine proteases the basicity of a histidine and the nucleophilicity of a serine, together with an aspatate residue belonging to the catalytic triad are of great importance (Baeten *et al.*, 1998). The oxyanion region (Asn residue of the oxyanion hole) also an important characteristic of all subtilisin-like serine protease (Siezen & Leunissen, 1997), was found in *Shewanella arctica* subtlisin-like protease (Asn-299) as well as in the subtilisins-like protease form the bacterial strains *Pseudoalteromonas* sp.(Asn-281), *Vibrio cholerae* (Asn-297), *Alteromonas* sp.(Asn-358), *Streplomyces albogriseolus* (Asn-3389, *Xanthomonas campesTris* (Asn-338), *Xanthomonas axonopodis* (Asn-338), *Fervidobacterium isladicum* (Asn-390), and from archaea *Thermococcus kadaeaensis* (Asn-182) (Gödde *et al.*, 2005; Kannan *et al.*, 2001; Lee *et al.*, 2000). At the N-terminal region of *Shewanella arctica* subtilisin-like protease a subtilisin-N conserved domain was found between the amino acids residues Asp-55 and Thr-135, and showed high similarity to the N-terminal subtlisin domains of the subtilisin-like protease from *Alteromonas* sp (Asp-36 to Thr-116), *Vibrio cholerae* (Asn-51 to Asn-131), *Vibrio alginolyticus* (Asp-56 to Asp-135), *Deinococcus radiodurans* (Gly-88 to Ser-175), *Moraxlla* sp. (Gly-30 to Ser-109) and *Bacillus* sp. (Lys-9 to Thr-78) (GenBank).

The new subtilisin-like serine protease form *Shewanella arctica* exhibits 65 % amino acid identity with the extracellular alkaline subtilisin-like serine protease from *Pseudoalteromonas* sp., 57 % identity with the subtilisin-like serine protease from *Alteromonas* sp., 44 % identity with alkaline subtilisin-like serine protease from *Vibrio cholerae*, and 36 % serine protease from the fungi *Arthrobotrys oligospora*. The phylogenetic analysis of *Shewanella arctica* subtilisin-like serine protease placed the enzyme next to the subtilisin-like serine protease from *Pseudoalteromonas* sp. The recombinant enzyme was purified in one step by Q-sepharose chromatography. The enzyme was purified 6.93-fold at yield of 8.67 % recovery and increase of specific activity from 0.075 U/mg to 0.52 U/mg. SDS-PAGE anaylysis showed that the molecular mass of the recombinant enzyme is 67.4 kDa and matches the predicted molecular mass from the primary structure. The 67.4 kDa of *Shewanella arctica* subtisilin-like serine protease is very close to the values reported for *Pseudoalteromonas* sp. subtilisin-like serine protease (65.1 kDa) (Lee *et al.*, 2000). Values from 33 to 115 kDa have been reported for subtilisin-like serine protease of the followed bacteria, archaea and fungi: *Pseudoalteromonas* sp., *Shewanella* sp. (strain

AC10), *Streplomyces albogriseolus*, *Mycobacterium tuberculosis*, *Xenorhabdus nemotophila*, *Thermoanaerobacter yonseiensis*, *Bacillus halmopalus*, *Fervidobacterium pennivorans*, *Fervidobacterium isladicum*, *Thermococcus kadaeaensis*, *Aeropyrum pernix*, *Natrialba magadii* and *Pleurotus ostreatus* (Caldas *et al.*, 2002; Dave *et al.*, 2002; Gimenez *et al.*, 2000; Gödde *et al.*, 2005; Jang *et al.*, 2002; Kannan *et al.*, 2001; Kluskens *et al.*, 2002; Kulakova *et al.*, 1999; Palmieri *et al.*, 2001; Saeki *et al.*, 2002; Suzuki *et al.*, 1997).

Serine proteases are generally active at neutral and alkaline pH, with optimal activity at pH 7 to 12, although higher pH optimum (10-12.5) have been detected from *Bacillus* species such as *Bacillus halmopalus* (pH 11-12) (Saeki *et al.*, 2002). The optimum pH of *Shewanella arctica* subtilisin-like serine protease (pH 8) is within the range of optimum pH reported for this protease family. The same pH optimum was reported for subtilisin-like serine protease from *Fervidobacterium isladicum*, archaea *Natrialba magadii*, and *Aeropyrum pernix* (Catara *et al.*, 2003; Gimenez *et al.*, 2000; Gödde *et al.*, 2005). Neutral and alkaline pH optimum was also reported for other several sublilisin-like serine protease form different bacteria and archaea such as: *Shewanella* sp. (strain AC10) (pH 9), *Streplomyces albogriseolus* (pH 10), *Xenorhabdus nemotophila* (pH 7), *Thermoanaerobacter yonseiensis* (pH 9), and *Thermococcus kadaeaensis* (pH 9.5) (Caldas *et al.*, 2002; Jang *et al.*, 2002; Kannan *et al.*, 2001; Kulakova *et al.*, 1999; Suzuki *et al.*, 1997). The temperature optimum (60 °C) for *Shewanella arctica* enzyme is also similar to the subtilisin-like serine protease from *Bacillus halmopalus* and *Aeropyrum pernix* (Catara *et al.*, 2003; Saeki *et al.*, 2002). Higher and lower optimal temperatures (20-92 °C) for subtilisin-like serine protease were reported. Optimal temperatures of 80 to 92 °C for subtilisin-like serine protease from *Streplomyces albogriseolus, Thermoanaerobacter yonseiensis,* and the archaea *Aeropyrum pernix* (Catara *et al.*, 2003; Jang *et al.*, 2002; Suzuki *et al.*, 1997) are some examples of thermoactive subtilisin-like serine protease reported. On the other hand, several subtilisin-like serine protease are reported to work at ranges between 20-30 °C, such as the subtilisins-like serine protease from: *Pseudoalteromonas* sp., *Shewanella* sp. (strain AC10), and *Xenorhabdus nemotophila* (Caldas *et al.*, 2002; Kulakova *et al.*, 1999; Lee *et al.*, 2000). The purified enzyme showed stability at 40 °C for more then 45 h and exhibited a half-life of 1.5 h at 70 °C. The remarkable ability of the *Shewanella arctica* subtilisin-like serine protease to work at a very wide range of temperatures (0 to 80 °C) makes this enzyme an attractive candidate in industrial processes running at elevated temperature varying the temperature.

Shewanella arctica subtilisin-like serine protease was completely inhibited by PSMF, chymostatin and antipain. E-64 and pepstalin partially inhibit the enzyme. PMSF, chymostatin and antipain are known to be serine protease inhibitors which react with the active-site serine residues. These three inhibitors inhibits most, if not all, subtilisin-like serine proteases reported to date (Caldas *et al.*, 2002; Catara *et al.*, 2003; Dave *et al.*, 2002; Gimenez *et al.*, 2000; Gödde *et al.*, 2005; Jang *et al.*, 2002; Lee *et al.*, 2000; Palmieri *et al.*, 2001; Suzuki *et al.*, 1997).

According to all data presented, the new *Shewanella arctica* protease is classified as a subtilisin-like serine protease. In conclusion, a novel subtilisin-like serine protease from *Shewanella arctica* sp. nov. was discovered (Patent registration number 102005028295.4). The enzyme was cloned and successfully expressed in *E. coli*. Maximal proteolytic activity was determined at 60 °C and pH 8. The enzyme was active between 0 to 80 °C.

5 SUMMARY

1- In the present study, two psychrophilic bacteria were isolated at 4 °C from sea water sample collected from Spitsbergen in the Arctic. One of the strains is strictly anaerobic and grows optimally at 10 °C and pH 7 in a medium containing 2 % NaCl (w/v), using pyruvate, starch, yeast extract, glucerol, xylose or 3-hydroxybutyrate as substrates. The fermentation products are butyrate, acetate and propionate. The fatty acid methyl esters (FAMEs) are composed of 28.17 % straight chain saturated FAMEs, 3.21 terminally branched saturated FAMEs and 22.13 % monounsaturated FAMEs. Phylogenetic analysis revealed a close relationship of the new isolate to *Ilyobacter polytropus*, with a 16S rDNA sequence identity of 92 % and a G+C value of 28.7 mol. This psychrophilic strain constitutes a novel species of the genus *Ilyobacter* and the name *Ilyobacter psychrophilus* is proposed.

A second psychrophilic, strictly aerobic bacterium was isolated and characterized. The strain grows optimally over the temperature range of 10-15 °C and a pH range of 7-8 in media containing 8 to 9 % NaCl (w/v), using various carbohydrate and organic acids as substrates. The fatty acid methyl esters (FAMEs) are composed of: 17.89 % straight chain saturated FAMEs, 14.85 % terminally branched saturated FAMEs and 17.73 % monounsaturated FAMEs. Phylogenetic analysis revealed a close relationship of the new isolate to *Shewanella putrefaciens* with 99 % 16S rDNA sequence identity, 50 % DNA-DNA similarity and G+C value of 48 % mol. Phylogenetic evidence, together with phenotypic characteristics, show that this isolate constitutes a novel species of the genus *Shewanella* and the name *Shewanella arctica* is proposed; DSM 16509.

2- A gene encoding a type I pullulanase was identified sequenced from a gene library of psychrophilic bacterium *Shewanella arctica* sp. nov., and actively expressed in *E coli* XLOLR. The gene 4.323 kb long, has a G C content of 47.97 % and encodes an enzyme with molecular mass of 155.9 kDa (1440 amino acids). The amino acid sequence indicated that the enzyme has a signal sequence of 23 amino acids (2.3 kDa). A Shine-Dalgarno-like sequence (AGGAGA) is found at 9 to 14 pb upstream from the start codon. The protein contains the seven-residue motif which is typical of type I pullulanases (FNWGYDP). In addition, the four regions conserved in amylolytic enzymes belonging to glycosyl hydrolysis family 13 and typically found in pullulanase were found. BLAST and phylogenitic analysis placed the enzyme next to the type I pullulanase from *Microbulbifer degradans* with 53 % amino acid identity. The native and recombinant enzymes were purified to a final specific activity of 3.0 U/mg and 66.6 U/mg respectively. The native pullulanase is monomeric enzyme with an apparent molecular mass of 155 kDa. Two recombinant pullulanases were detected on SDS-gel with apparent molecular masses of 85 and 70 kDa. Native and recombinant enzymes are active at a broad range of temperatures (10 – 50 °C for native and 4 – 60 °C for recombinant enzyme) and pH (4 – 9 for native and 6 – 8 for recombinant enzymes). Showing optimum pullulytic activity at 45 °C and pH 7. The two recombinant enzymes are stable at 40 °C for 3h and exhibited a half-life of 44 min at 50 °C. The hydrolysis of pullulan revealed the complete conversion of pullulan to maltotriose. Mn^{+2}, Zn^{+2}, Cu^{+2}, Fe^{+2}, EDTA, sodium dodecysulfate, urea, dithiothreitol, guanidine-HCl and iodocetamide inhibited the pullulytic activity.

3- A gene encoding a subtilisin-like serine protease was identified and sequenced from a gene library of psychrophilic bacterium *Shewanella arctica* sp. nov.. The gene was amplified by PCR, cloned and activity expressed in *E. coli* Tuner (DE3) pLacl. The gene 1.935 kb long, has a CG content of 48.78 % and encodes an enzyme with molecular mass of 67.4 kDa (644 amino acids). The amino acid sequence indicated that the enzyme has a signal sequence of 38 amino acids (4.4 kDa). The protein contains the three catalytic residues (Asp-180, His-213 and Ser-364) and an oxyanion region (Asn-299 residue of the oxyanion hole). At the N-terminal region a subtilisin-N domain was detected between Asp-55 and Thr-135 residues. Amino acids sequence comparison and phylogenetic analysis indicated that this enzyme can be classified as subtilisin-like serine protease and placed it next to the subtilisin-like serine protease form *Pseudoalteromonas* sp. with 65 % amino acids identity. The recombinant subtilisin-like serine protease was purified to a specific activity of 0.52 U/mg with Q-sepharose chromatography. The enzyme is active at broad range of temperatures (0 – 80 °C) and pH (6 – 10), showing optimal proteolytic activity at 60 °C and pH 8. The enzyme is stable at 40 °C for 45h and exhibited a half-life of 1.5h at 70 °C. The enzyme is inhibited by chymostatin, phenylmethylsulfonyl fluoride (PMSF) and antipain-dehydrochoride.

6 ZUSAMMENFASSUNG

1- In der vorliegenden Studie wurden zwei neue, psychrophile Bakterienspezies von Spitsbergen stammenden arktischen Seewasserproben bei 4 °C isolier. Einer der Stämme war strikt anaerob, wuchs optimal bei 10°C und pH 7 in einen 2% NaCl (w/v) enthaltendem Medium mit Pyruvat, Stärke, Hefeextrakt, Glucose, Xylose oder 3-Hydroxybutyrat als Substrat. Als Fermentationsprodukte wurden Butyrat, Acetat und Propionat gebildet. Die Fettsäureanalyse ergab eine Zusammensetzung von 28,2% linearen, gesättigten Fettsäuren, 3,2% endständig verzweigten, gesättigten Fettsäuren und 22,1% einfach ungesättigten Fettsäuren. Die phylogenetische Analyse zeigte mit einer 16S rDNA Sequenzidentität von 92% und einem GC-Gehalt von 28.7 Molprozent eine enge Verwandtschaft des Neuisolats zu *Ilyobacter polytropus*. Zusammen mit den phänotypischen Charakteristika beweist dies, dass es sich bei diesem psychrophilen Stamm um eine neue Art der Gattung *Ilyobacter* handelt, für die der Name *Ilyobacter psychrophilus* vorgeschlagen wird. Ein zweites, pyschrophiles, strikt aerobes Bacterium wurde isoliert und charakterisiert. Der Stamm wuchs optimal in einem Temperaturbereich von 10-15°C, einem pH-Bereich von 7-8 in einem Medium mit 8 bis 9 % NaCl (w/v) und mit einer Vielzahl von Kohlenhydraten und organische Saüren als Substrat. Die Fettsäuren setzten sich aus 17,9% linearen gesättigten Fettsäuren, 14,9% endständig verzweigten gesättigten Fettsäuren und 17,7% einfach ungesättigten Fettsäuren zusammen. Die phylogenetische Analyse ergab mit einer 99%igen 16S rDNA Sequenzidentität, einer 50%igen DNA-DNA Ähnlichkeit und einem GC-Gehalt von 48 Molprozent eine nahe

Verwandtschaft zu dem neuisolierten *Shewanella putrefaciens*. Zusammen mit den phänotypischen Charakteristika beweist diese Analyse, dass es sich bei diesem psychrophilen Stamm um eine neue Art der Gattung *Shewanella* handelt. Bezugnehmend auf den Isolationsort wird der Name *Shewanella arctica* vorgeschlagen; DSMZ 16509.

2- Ein für eine Pullulanase des Typs I kodierendes Gen wurde in einer Genbank von *Shewanella arctica* sp. nov. identifiziert, sequenziert und aktiv in *E. coli* XLOLR exprimiert. Das Gen bestand aus 4323 bp, besaß einen GC-Gehalt von 48% und kodiert für ein Enzym mit einem errechneten Molekulargewicht von 155,9 kDa (1440 Aminosäuren). Das Enzym besitzt eine putative, aus 23 Aminosäuren (2,3 kDa) bestehende Signalsequenz. Eine Shine-Dalgarno-ähnliche Sequenz (AGGAGA) zeigte sich 9 bis 14 pb upstream des Startcodons. Das Enzym wies das für Pullulanasen des Typ I typische, aus 7 Aminosäuren bestehende Motiv (FNWGYDP) und die vier konservierten Bereiche der Enzyme der Glycosylhydrolase-Familie 13 auf. BLAST und phylogenetische Analysen ergaben mit einer Aminosäure-Identität von 53% die höchste Ähnlichkeit zur Typ I-Pullulanase aus *Microbulbifer degradans*. Das native Enzym wurde zu einer spezifischen Aktivität von 3.0 U/mg und das rekombinante Enzym zu einer spezifischen Aktivität von 66.6 U/mg aufgereinigt. Die native Pullulanase ist ein monomeres Enzym mit einem Molekulargewicht von 155kDa. Im SDS-Gel konnten zwei rekombiante Pullulanasen mit Molekulargewichten von 85 und 70 kDa nachgewiesen werden. Natives und rekombinantes Enzym waren aktiv in einem breiten Temperaturbereich (10–50°C für das native und 4 – 60 °C für das rekombinante Enzym) und pH-Bereich (4 – 9 für das native und 6 – 8 für das rekombinante Enzym). Die optimale Aktivität lag bei 45°C und pH 7 vor. Das rekombinante Enzym war bei 40°C für 3 h stabil und wies bei 50°C eine Halbwertzeit von 44 min auf. HPLC-Analysen zeigten, dass das Enzym Pullulanan komplett zu Maltotriose umsetzte. Mn^{+2}, Zn^{+2}, Cu^{+2}, Fe^{+2}, EDTA, SDS, Harnstoff, Dithiothreitol, Guanidine-HCl und Iodocetamide hemmten die Pullulanase.

3- Ein für eine Subtilisin-ähnliche Serinprotease kodierendes Gen wurde in einer Genbank von *Shewanella arctica* sp. nov. identifiziert und sequenziert. Das Gen wurde mittels PCR amplifiziert, kloniert und aktiv in *E. coli* Tuner (DE3) pLacl exprimiert. Das Gen bestand aus 1935 bp, hatte ein GC-Gehalt von 48,8% und kodiert für ein Enzym mit einem Molekulargewicht von 67,4 kDa (644 Aminosäuren). Das Enzym besitzt eine putative, aus 38 Aminosäuren bestehende Signalsequenz (4,4 kDa). Das Protein enthält die drei katalytischen Reste (Asp-180, His-213 and Ser-364) und die Oxyanionregion (Asn-299). N-terminal wurde zwischen Asp-55 und Thr-135 eine Subtilisin-N-Domäne detektiert. Aminosäuresequenzvergleiche und phylogenetische Analysen legen nahe, dass das Enzyme als Subtilisin ähnliche Serinprotease klassifiziert werden kann. Die höchste Sequenzidentität von 65% wurde zur Subtilisin-ähnlichen Serinprotease von *Pseudoalteromonas* sp. gefunden. Das rekombinante Subtilisin wurde zu einer spezifischen Aktivität von 0.52 U/mg aufgereinigt. Das Enzym war aktiv in einem breiten Temperatur- (0 – 80 °C) und pH-Bereich (pH 6 – 10). Optimal aktiv war das Enzym bei 60°C und pH 8. Das Enzym war bei 40°C für 45h stabil und wies bei 70°C eine Halbwertzeit von 1,5 h auf. Das Enzym wurde durch Chymostatin, Phenylmethylsulfonylfluoride (PMSF) und Antipain-dehydrochloride inhibiert.

7 REFERENCES

Abraham, L. D. & Breuil, C. (1995). Factors affecting autolysis of a subtilisin-like serine proteinase secreted by *Ophiostoma piceae* and identification of the cleavage site. *Biochim Biophys Acta* **1245**, 76-84.

Aghajari, N., Feller, G., Gerday, C. & Haser, R. (1998). Structures of the psychrophilic *Alteromonas haloplanctis* alpha-amylase give insights into cold adaptation at a molecular level. *Structure* **6**, 1503-1516.

Albertson, G. D., Mchale, R. H., Gibbs, M. D. & Bergquist, P. L. (1997). Cloning and sequence of a type I pullulanase from an extremely thermophilic anaerobic bacterium, *Caldicellulosiruptor saccharolyticus*. *Biochim Biophys Acta* **1354**, 35-39.

Altschul, S. F., Gish, W., Miller, W., Myers, E. W. & Lipman, D. J. (1990). Basic local alignment search tool. *J Mol Biol* **215**, 403-410.

Amemura, A., Chakraborty, R., Fujita, M., Noumi, T. & Futai, M. (1988). Cloning and nucleotide sequence of the isoamylase gene from *Pseudomonas amyloderamosa* SB-15. *J Biol Chem* **263**, 9271-9215.

Antranikian, G., Vorgias, C. E. & Bertoldo, C. (2005). Extreme environmental as a resource for microorganisms and novel biocatalysts. *Adv Biochem Eng Biotechnol* **96**,

Argos, P. (1987). A sensitive procedure to compare amino acid sequences. *J. Mol. Biol.* **193**, 385-396.

Baeten, A., Maes, D. & Geerlings, P. (1998). Quantumchemical study of the catalytic triad in subtilisin: the influence of amino acid substitutions on enzymatic activity. *J Theor Biol* **195**, 27-40.

Barett, A. J. (1994). Proteolytic enzymes: serine and cysteine peptidases. *Methods Enzymol.* **244**, 1-15.

Barett, A. J. (1995). Proteolytic enzymes: aspartic and metallopeptidases. *Methods Enzymol.* **248**, 183.

Barr, P. J. (1991). Mammalian subtilisins: the long-sought dibasic processing endoproteases. *Cell* **66**, 1-3.

Bernfeld, P. (1955). Amylase a and b. *Methods Enzymol* **1**, 149-155.

Berry, J. A. & Magoon, A. (1934). Growth of microorganisms at and below 0 °C. *Phytopathology* **24**, 780-796.

Bertoldo, C. & Antranikian, G. (2002a). Natural polysaccharide-degradation enzymes. In *Enzymes catalysis in organic synthesis*. Edited by K. Drauz & H. Waldmann. Wiley-VCH Verlag, Weinheim

Bertoldo, C. & Antranikian, G. (2002b). Starch-hydrolyzing enzymes from thermophilic archaea and bacteria. *Curr Opin Chem Biol* **6**, 151-160.

Bertoldo, C., Armbrecht, M., Becker, F., Schafer, T., Antranikian, G. & Liebl, W. (2004). Cloning, sequencing, and characterization of a heat- and alkali-stable type I pullulanase from *Anaerobranca gottschalkii*. *Appl Environ Microbiol* **70**, 3407-3416.

Bertoldo, C., Duffner, F., Jorgensen, P. L. & Antranikian, G. (1999). Pullulanase type I from *Fervidobacterium pennavorans* Ven5: cloning, sequencing, and expression of the gene and biochemical characterization of the recombinant enzyme. *Appl Environ Microbiol* **65**, 2084-2091.

Bibel, M., Brettl, C., Gosslar, U., Kriegshauser, G. & Liebl, W. (1998). Isolation and analysis of genes for amylolytic enzymes of the hyperthermophilic bacterium *Thermotoga maritima*. *FEMS Microbiol Lett* **158**, 9-15.

Bligh, E. G. & Dyer, W. J. (1959). A rapid method of total lipid extraction and purification. *Can J Biochem Physiol* **37**, 911-7.

Bowman, J. P., Mccammon, S. A., Nichols, D. S., Skerratt, J. H., Rea, S. M., Nichols, P. D. & Mcmeekin, T. A. (1997). *Shewanella gelidimarina* sp. nov. and *Shewanella frigidimarina* sp. nov., novel Antarctic species with the ability to produce eicosapentaenoic acid (20:5 omega 3) and grow anaerobically by dissimilatory Fe(III) reduction. *Int J Syst Bacteriol* **47**, 1040-1047.

Bozal, N., Montes, M. J., Tudela, E., Jimenez, F. & Guinea, J. (2002). *Shewanella frigidimarina* and Shewanella *livingstonensis* sp. nov. isolated from Antarctic coastal areas. *Int J Syst Evol Microbiol* **52**, 195-205.

Bradford, M. M. (1976). A rapid and sensitive method for the quantitation of microgram quantities of protein utilizing the principle of protein-dye binding. *Anal Biochem* **72**, 248-254.

Brenner, S. (1988). The molecular evolution of genes and proteins: a tale of two serines. *Nature* **334**, 528-30.

Brettar, I., Christen, R. & Hofle, M. G. (2002). *Shewanella denitrificans* sp. nov., a vigorously denitrifying bacterium isolated from the oxic-anoxic interface of the Gotland Deep in the central Baltic Sea. *Int J Syst Evol Microbiol* **52**, 2211-2217.

Britton, H. T. S. & Robinson, R. E. (1931). Universal buffer solution and dissociation constant of veronal. *J. Chem. Soc.* 1456-1473.

Brown, G. M., Campbell, I. & Priest, F. G. (1988). Introduction to biotrchnology. Edited by English language book society / Blackwell scientific puplications, UK

Brune, A., Evers, S., Kaim, G., Ludwig, W. & Schink, B. (2002). *Ilyobacter insuetus* sp. nov., a fermentative bacterium specialized in the degradation of hydroaromatic compounds. *Int J Syst Evol Microbiol* **52**, 429-432.

Buchholz-Cleven, B. E. E., Rattunde, B. & Staub, K. L. (1997). Screening forgenetic diversity of anaerobic Fe(II)- oxidizing bacteria using DGGE and whole-cell hybridization. *Syst Appl Microbiol* **20**, 301.309.

Bull, A. T., Holt, J. G. & Lilly, M. D. (1982). Biotechnology. International trends and perspectives. Edited by Organization for economic cooperation and development, Paris

Bu'lock, J. & Kristiansen (1987). Basic biotechnology. Edited by H. Brace. Jovanovich, London

Burton, K. S., Wood, D. A., Thurston, C. F. & Barker, P. J. (1993). Purification and characterization of a serine proteinase from senescent sporophores of the commercial mushroom Agaricus bisporus. *J Gen Microbiol* **139 Pt 6**, 1379-86.

Caldas, C., Cherqui, A., Pereira, A. & Simoes, N. (2002). Purification and characterization of an extracellular protease from *Xenorhabdus nematophila* involved in insect immunosuppression. *Appl Environ Microbiol* **68**, 1297-1304.

Carter, P. & Wells, J. A. (1988). Dissecting the catalytic triad of a serine protease. *Nature* **332**, 564-568.

Cashion, P., Holder-Franklin, M. A., Mccully, J. & Franklin, M. (1977). A rapid method for the base ratio determination of bacterial DNA. *Anal Biochem* **81**, 461-6.

Catara, G., Ruggiero, G., La Cara, F., Digilio, F. A., Capasso, A. & Rossi, M. (2003). A novel extracellular subtilisin-like protease from the hyperthermophile *Aeropyrum pernix* K1: biochemical properties, cloning, and expression. *Extremophiles* **7**, 391-399.

Chestukhina, G. G., Zalunin, I. A., Kostina, L. I., Kotova, T. S., Kattrukha, S. P. & Stepanov, V. M. (1980). Crystal-forming proteins of *Bacillus thuringiensis*. The limited hydrolysis by endogeneous proteinases as a cause of their apparent multiplicity. *Biochem J* **187**, 457-465.

Dave, J. A., Gey Van Pittius, N. C., Beyers, A. D., Ehlers, M. R. & Brown, G. D. (2002). Mycosin-1, a subtilisin-like serine protease of *Mycobacterium tuberculosis*, is cell wall-associated and expressed during infection of macrophages. *BMC Microbiol* **2**, 30-38.

David, P. (1999). The fifth miracle. Edited by Simon & Schuster. New York

De Ley, J., Cattoir, H. & Reynaerts, A. (1970). The quantitative measurement of DNA hybridization from renaturation rates. *Eur J Biochem* **12**, 133-142.

Dong, G., Vieille, C. & Zeikus, J. G. (1997). Cloning, sequencing, and expression of the gene encoding amylopullulanase from Pyrococcus furiosus and biochemical characterization of the recombinant enzyme. *Appl Environ Microbiol* **63**, 3577-84.

Duffner, F., Bertoldo, C., Andersen, J. T., Wagner, K. & Antranikian, G. (2000). A new thermoactive pullulanase from *Desulfurococcus mucosus*: cloning, sequencing, purification, and characterization of the recombinant enzyme after expression in Bacillus subtilis. *J Bacteriol* **182**, 6331-6338.

Durham, D. R. (1993). The elastolytic properties of subtilisin GX from alkalophilic *Bacillus* sp. strain 6644 provides a means of differentiation from other subtilisins. *Biochem Biophys Res Commun* **194**, 1365-1370.

Erra-Pujada, M., Deleire, F., Duchiron, F. & O'Donohue, M. J. (1999). The type II pullulanase of Thermococcus hydrothermalis: molecular characterization of the gene and expression of the catalytic domain. *J. Bacteriol.* **181**, 3282-3287.

Felsenstien, J. (1995). PHYLIP (Phylogeny Inference Package), version 3.57c. Edited by Department of Genetics, University of Washington, Seattle, WA, USA

Freeman, S. A., Peek, K., Prescott, M. & Daniel, R. (1993). Characterization of a chelator-resistant proteinase from *Thermus* strain Rt4A2. *Biochem J* **295**, 463-469.

Garrity, G. M. & Holt, J. G. (2001). The road map to the Manual. In *Berger's Manual of Systematic Bacteriology*, 2nd edn, vol. 1, pp 119-166. Edited by D. R. Boone & R. W. Castenholz. New York: Springer

Gaucher, G. M. & Stevenson, K. J. (1976). Thermomycolin. *Methods Enzymol* **45**, 415-33.

Gauthier, G., Gauthier, M. & Christen, R. (1995). Phylogenetic analysis of the genera *Alteromonas*, *Shewanella*, and *Moritella* using genes coding for small-subunit rRNA sequences and division of the genus *Alteromonas* into two genera, *Alteromonas* (emended) and *Pseudoalteromonas* gen. nov., and proposal of twelve new species combinations. *Int J Syst Bacteriol* **45**, 755-761.

Gerday, C., Aittaleb, M., Bentahir, M., Chessa, J. P., Claverie, P., Collins, T., D'amico, S., Dumont, J., Garsoux, G., Georlette, D.,

Hoyoux, A., Lonhienne, T., Meuwis, M. A. & Feller, G. (2000). Cold-adapted enzymes: from fundamentals to biotechnology. *Trends Biotechnol* **18**, 103-7.

Gerhardt, P., Murray, R. G. E., Wood, W. A. & Krieg, N. R. (1994). *Method of General and Molecular Microbiology.* Edited by P. Gerhardt, R. G. E. Murray, W. A. Wood & N. R. Krieg. Washington, DC: American Society for Microbiology

Gianese, G., Argos, P. & Pascarella, S. (2001). Structural adaptation of enzymes to low temperatures. *Protein Eng* **14**, 141-8.

Gimenez, M. I., Studdert, C. A., Sanchez, J. J. & De Castro, R. E. (2000). Extracellular protease of *Natrialba magadii*: purification and biochemical characterization. *Extremophiles* **4**, 181-188.

Gödde, C., Sahm, K., Brouns, S. J. J., Kluskens, L. D., Van Der Oost, J., M. De Vos, W. & Antranikian, G. (2005). Islandisin, a new thermostsble subtilisin from *Fervidobacterium islandicum*: cloning and expression in *E. coli. Appl Environ Microbiol* **7**,

Gomes, J. & Steiner, W. (2004). The biocatalytic potential of extremophiles and extremozymes. *Food Technol. Biotechnol.* **42**, 223-235.

Gounot, A. M. (1999). Microbial life in permanently cold soils. In *Cold-adapted Organisms: Ecology, Physiology, Enzymology and Molecular Biology*, pp. 3-17. Edited by R. Margesin & F. Schinner. Heidelberg: Springer-Verlag

Grant, W. D. (1991). General view of halophiles. In *Superbugs-Microorganisms in extreme environments*. Edited by K. Horikoshi & W. D. Grant. Spinger-verlag, Tokyo

Gupta, R., Beg, Q. K. & Lorenz, P. (2002). Bacterial alkaline protease: molecular approaches and industrial applications. *Appl Microbiol Biotechnol* **59**, 15-32.

Hacking, A. J. (1987). Economic aspects of biotechnology, Cambridge studies in biotechnology. Edited by Cambridge university press, Cambridge

Hartley, B. S. (1960). Proteolytic enzymes. *Annu Rev Biochem* **29**, 45-72.

Hough, D. W. & Danson, M. J. (1999). Extremozymes. *Curr Opin Chem Biol* **3**, 39-46.

Huss, V. A. R., Festl, H. & Schleifer, K. H. (1983). Studies on the spectrophotometric determination od DND hybridization from renaturation rates. *Syst Appl Microbiol* **4**, 184-192.

Ingraham, J. L. & Stokes, J. L. (1959). Psychrophilic bacteria. *Bacteriol Rev* **23**, 97-108.

Ivanova, E. P., Gorshkova, N. M., Bowman, J. P., Lysenko, A. M., Zhukova, N. V., Sergeev, A. F., Mikhailov, V. V. & Nicolau, D. V. (2004a). *Shewanella pacifica* sp. nov., a polyunsaturated fatty acid-producing bacterium isolated from sea water. *Int J Syst Evol Microbiol* **54**, 1083-1087.

Ivanova, E. P., Nedashkovskaya, O. I., Sawabe, T., Zhukova, N. V., Frolova, G. M., Nicolau, D. V., Mikhailov, V. V. & Bowman, J. P. (2004b). *Shewanella affinis* sp. nov., isolated from marine invertebrates. *Int J Syst Evol Microbiol* **54**, 1089-1093.

Ivanova, E. P., Nedashkovskaya, O. I., Zhukova, N. V., Nicolau, D. V., Christen, R. & Mikhailov, V. V. (2003a). *Shewanella waksmanii* sp. nov., isolated from a sipuncula (*Phascolosoma japonicum*). *Int J Syst Evol Microbiol* **53**, 1471-1477.

Ivanova, E. P., Sawabe, T., Gorshkova, N. M., Svetashev, V. I., Mikhailov, V. V., Nicolau, D. V. & Christen, R. (2001). *Shewanella japonica* sp. nov. *Int J Syst Evol Microbiol* **51**, 1027-33.

Ivanova, E. P., Sawabe, T., Hayashi, K., Gorshkova, N. M., Zhukova, N. V., Nedashkovskaya, O. I., Mikhailov, V. V., Nicolau, D. V. & Christen, R. (2003b). *Shewanella fidelis* sp. nov., isolated from sediments and sea water. *Int J Syst Evol Microbiol* **53**, 577-582.

Ivanova, E. P., Sawabe, T., Zhukova, N. V., Gorshkova, N. M., Nedashkovskaya, O. I., Hayashi, K., Frolova, G. M., Sergeev, A. F., Pavel, K. G., Mikhailov, V. V. & Nicolau, D. V. (2003c). Occurrence and diversity of mesophilic *Shewanella strains* isolated from the North-West Pacific Ocean. *Syst Appl Microbiol* **26**, 293-301.

Jahnke, K. D. (1992). Basic computer for evaluation of spectoscopic DNA renaturation data from GILFORD System 2600 spectrometer on a PC/XT/AT type personal computer. *J Microbiol Methods* **15**, 61-73.

Jang, H. J., Kim, B. C., Pyun, Y. R. & Kim, Y. S. (2002). A novel subtilisin-like serine protease from *Thermoanaerobacter yonseiensis* KB-1: its cloning, expression, and biochemical properties. *Extremophiles* **6**, 233-243.

Janssen, P. H. & Liesack, W. (1995). Succinate decarboxylation by *Propionigenium maris* sp. nov., a new anaerobic bacterium from an estuarine sediment. *Arch Microbiol* **164**, 29-35.

Kalisz, H. M. (1988). Microbial proteinases. *Adv Biochem Eng Biotechnol* **36**, 1-65.

Kannan, Y., Koga, Y., Inoue, Y., Haruki, M., Takagi, M., Imanaka, T., Morikawa, M. & Kanaya, S. (2001). Active subtilisin-like protease from a hyperthermophilic archaeon in a form with a putative prosequence. *Appl Environ Microbiol* **67**, 2445-2452.

Katsuya, Y., Mezaki, Y., Kubota, M. & Matsuura, Y. (1998). Three-dimensional structure of *Pseudomonas* isoamylase at 2.2 A resolution. *J. Mol. Biol.* **281**, 885-897.

Kelly, A. P., Diderichsen, B., Jorgensen, S. & Mcconnell, D. J. (1994). Molecular genetic analysis of the pullulanase B gene of *Bacillus acidopullulyticus*. *FEMS Microbiol Lett* **115**, 97-105.

Kim, C. H., Nashiru, O. & Ko, J. H. (1996). Purification and biochemical characterization of pullulanase type I from *Thermus caldophilus* GK-24. *FEMS Microbiol Lett* **138**, 147-152.

Kimura, M. (1980). A simple method for estimating evolutionary rates of base substitutions through comparative studies of nucleotide sequences. *J Mol Evol* **16**, 111-20.

Kistjansson, J. K. & Herggvidsson, G. O. (1995). Ecology and habitats of extremophiles. *World J. Microbiol. Biotecnol.* **11**, 17-25.

Kluskens, L. D., Voorhorst, W. G., Siezen, R. J., Schwerdtfeger, R. M., Antranikian, G., Van Der Oost, J. & De Vos, W. M. (2002). Molecular characterization of fervidolysin, a subtilisin-like serine protease from the thermophilic bacterium *Fervidobacterium pennivorans*. *Extremophiles* **6**, 185-194.

Kohn, G., Van Der Ploeg, P., Mobius, M. & Sawatzki, G. (1996). Influence of the derivatization procedure on the results of the gaschromatographic fatty acid analysis of human milk and infant formulae. *Z Ernahrungswiss* **35**, 226-34.

Kornacker, M. G. & Pugsley, A. P. (1990). Molecular characterization of pulA and its product, pullulanase, a secreted enzyme of *Klebsiella pneumoniae* UNF5023. *Mol Microbiol* **4**, 73-85.

Kujawski, M., Ziobro, R. & Gambus, H. (2002). Raw starch degradation by pullulanase. *Technologia Alimentaria* **1**, 31-35.

Kulakova, L., Galkin, A., Kurihara, T., Yoshimura, T. & Esaki, N. (1999). Cold-active serine alkaline protease from the psychrotrophic bacterium *Shewanella* strain ac10: gene cloning and enzyme purification and characterization. *Appl Environ Microbiol* **65**, 611-617.

Kumar, C. G. & Takagi, H. (1999). Microbial alkaline proteases: from a bioindustrial viewpoint. *Biotechnol Adv* **17**, 561-94.

Kunitate, A., Okamoto, M. & Ohmori, I. (1989). Purification and characterization of a thermostable serine protease from *Bacillus thuringiensis*. *Argric. Biol. Chem* **53**, 3251-3256.

Kunitz, M. (1947). Crystalline soybean trypsin inhibitor II. General properties. *J. Gen. Physiol.* **30**, 291-310.

Kuriki, T. & Imanaka, T. (1999). The concept od the a-amylase family: Structural similarty and common catalytic mechanism. *J Biosci Bioeng* **87**, 557-565.

Kuriki, T., Okada, S. & Imanaka, T. (1988). New type of pullulanase from *Bacillus stearothermophilus* and molecular cloning and expression of the gene in *Bacillus subtilis*. *J Bacteriol* **170**, 1554-1559.

Kwon, Y. T., Kim, J. O., Moon, S. Y., Yoo, Y. D. & Rho, H. M. (1995). Cloning and characterization of the gene encoding an extracellular alkaline serine protease from *Vibrio metschnikovii* strain RH530. *Gene* **152**, 59-63.

Laemmli, U. K. (1970). Cleavage of structural proteins during the assembly of the head of bacteriophage T4. *Nature* **227**, 680-685.

Larcher, G., Cimon, B., Symoens, F., Tronchin, G., Chabasse, D. & Bouchara, J. P. (1996). A 33 kDa serine proteinase from *Scedosporium apiospermum*. *Biochem J* **315**, 119-126.

Lee, S. O., Kato, J., Takiguchi, N., Kuroda, A., Ikeda, T., Mitsutani, A. & Ohtake, H. (2000). Involvement of an extracellular protease in algicidal activity of the marine bacterium *Pseudoalteromonas* sp. strain A28. *Appl Environ Microbiol* **66**, 4334-4339.

Leiros, H. K., Willassen, N. P. & Smalas, A. O. (2000). Structural comparison of psychrophilic and mesophilic trypsins. Elucidating the molecular basis of cold-adaptation. *Eur J Biochem* **267**, 1039-1049.

Leonardo, M. R., Moser, D. P., Barbieri, E., Brantner, C. A., Macgregor, B. J., Paster, B. J., Stackebrandt, E. & Nealson, K. H. (1999). *Shewanella pealeana* sp. nov., a member of the microbial community associated with the accessory nidamental gland of the squid Loligo pealei. *Int J Syst Bacteriol* **49**, 1341-1351.

Lepage, G. & Roy, C. C. (1984). Improved recovery of fatty acid through direct transesterification without prior extraction or purification. *J Lipid Res* **25**, 1391-6.

Macdonell, M. T. & Colwell, R. R. (1985). Pylogeny of the *Vibrionaceae*, and recommendation for two new genera, *Listonella* and *Shewanella*. *Syst Appl Microbiol* **6**, 171-182.

Macgregor, E. A. (2003). Limit dextrinase/Pullulanase. In *Handbook of food enzymology*. Edited by J. W. Whitaker, A. G. J. Voragen & D. W. S. Wong. Marcel dekker, Inc. New York

Madigan, M. T., Martinko, J. M. & Parker, J. (1997). Biology if microorganisms. Edited by M. T. Madigan, J. M. Martinko & J. Parker. Prentice hall international, USA

Madigan, M. T., Martinko, J. M. & Parker, J. (2000). Brok biology of microorganisms. Edited by P. F. Corey. Prentice Hall, Upper saddle river,NJ

Makemson, J. C., Fulayfil, N. R., Landry, W., Van Ert, L. M., Wimpee, C. F., Widder, E. A. & Case, J. F. (1997). *Shewanella woodyi* sp. nov., an exclusively respiratory luminous bacterium isolated from the Alboran Sea. *Int J Syst Bacteriol* **47**, 1034-1039.

Marhuenda-Egea, F. C., Piera-Velazquez, S., Cadenas, C. & Cadenas, E. (2002). An extreme halophilic enzyme active at low salt in reversed micelles. *J Biotechnol* **93**, 159-64.

Maruyama, A., Honda, D., Yamamoto, H., Kitamura, K. & Higashihara, T. (2000). Phylogenetic analysis of psychrophilic bacteria isolated from the Japan Trench, including a description of the deep-sea species *Psychrobacter pacificensis* sp. nov. *Int J Syst Evol Microbiol* **50**, 835-46.

Menon, A. S. & Goldberg, A. L. (1987). Protein substrates activate the ATP-dependent protease La by promoting nucleotide binding and release of bound ADP. *J Biol Chem* **262**, 14929-34.

Mesbah, M. & Whitman, W. B. (1989). Measurement of deoxyguanosine/thymidine ratios in complex mixtures by high-performance liquid chromatography for determination of the mole percentage guanine + cytosine of DNA. *J Chromatogr* **479**, 297-306.

Messoud, B. E., Ammar, Y. B., Mellouli, L. & Bejar, S. (2002). Thermostable pullulanase typ I from new isolated *Bacillus thermoleovorans* US105: cloning, sequencing and expression of the gene in *E. coli*. *Enzyme Microbiol. Technol.* **31**, 827-832.

Michaelis, S., Chapon, C., D'enfert, C., Pugsley, A. P. & Schwartz, M. (1985). Characterization and expression of the structural gene for pullulanase, a maltose-inducible secreted protein of *Klebsiella pneumoniae*. *J Bacteriol* **164**, 633-638.

Morita, R. Y. (1975). Psychrophilic bacteria. *Bacteriol Rev* **39**, 144-67.

Myers, C. R. & Nealson, K. H. (1988). Bacterial manganese reduction and growth with manganese oxide as the sole electron accepter. *Science* **240**, 1319-1321.

Nakajima, R., Imanaka, T. & Aiba, S. (1986). Comparison of amino acids sequences of eleven differents a-amylase. *Appl. Microbiol. Biotechnol.* **23**, 355-360.

Niehaus, F., Peters, A., Groudieva, T. & Antranikian, G. (2000). Cloning, expression and biochemical characterisation of a unique thermostable pullulan-hydrolysing enzyme from the hyperthermophilic archaeon Thermococcus aggregans. *FEMS Microbiol Lett* **190**, 223-9.

Nogi, Y., Kato, C. & Horikoshi, K. (1998). Taxonomic studies of deep-sea barophilic *Shewanella* strains and description of *Shewanella violacea* sp. nov. *Arch Microbiol* **170**, 331-338.

Palmieri, G., Bianco, C., Cennamo, G., Giardina, P., Marino, G., Monti, M. & Sannia, G. (2001). Purification, characterization, and functional role of a novel extracellular protease from *Pleurotus ostreatus*. *Appl Environ Microbiol* **67**, 2754-2759.

Petrovskis, E. A., Vogel, T. M. & Adriaens, P. (1994). Effects of electron acceptors and donors on transformation of tetrachloromethane by *Shewanella putrefaciens* MR-1. *FEMS Microbiol Lett* **121**, 357-363.

Plant, A. R., Morgan, H. W. & Daniel, R. M. (1986). A highly stable pullulanase from *Thermus aquaticus* YT-1. *Enzyme Microbiol. Technol.* **8**, 668-672.

Prentis, S. (1989). Biotechnology, A new industrial revolution. Edited by G. Braziller. New york

Prescott, L. M., P., P. J. & Klein, D. A. (1999). Microbiology. Edited by L. M. Prescott, P. J. P. & D. A. Klein. WCB McGraw-Hill, USA

Rainey, F. A. & Stackebrandt, E. (1993). 16S rDNA analysis reveals phylogenetic diversity among the polysaccharolytic clostridia. *FEMS Microbiol Lett* **113**, 125-128.

Rao, M. B., Tanksale, A. M., Ghatge, M. S. & Deshpande, V. V. (1998). Molecular and biotechnological aspects of microbial proteases. *Microbiol Mol Biol Rev* **62**, 597-635.

Rawlings, N. D. & Barrett, A. J. (1993). Evolutionary families of peptidases. *Biochem J* **290 (Pt 1)**, 205-18.

Rawlings, N. D. & Barrett, A. J. (1994). Families of serine peptidases. *Methods Enzymol.* **244**, 19-61.

Reid, G. A. & Gordon, E. H. (1999). Phylogeny of marine and freshwater *Shewanella*: reclassification of *Shewanella putrefaciens* NCIMB 400 as *Shewanella frigidimarina*. *Int J Syst Bacteriol* **49**, 189-191.

Rozzell, J. D. (1999). Commercial scale biocatalysis: myths and realities. *Bioorg Med Chem* **7**, 2253-61.

Rudiger, A., Jorgensen, P. L. & Antranikian, G. (1995). Isolation and characterization of a heat-stable pullulanase from the hyperthermophilic archaeon *Pyrococcus woesei* after cloning and expression of its gene in *Escherichia coli*. *Appl Environ Microbiol* **61**, 567-575.

Saeki, K., Hitomi, J., Okuda, M., Hatada, Y., Kageyama, Y., Takaiwa, M., Kubota, H., Hagihara, H., Kobayashi, T., Kawai, S. & Ito, S. (2002). A novel species of alkaliphilic *Bacillus* that produces an oxidatively stable alkaline serine protease. *Extremophiles* **6**, 65-72.

Satomi, M., Oikawa, H. & Yano, Y. (2003). *Shewanella marinintestina* sp. nov., *Shewanella schlegeliana* sp. nov. and *Shewanella sairae* sp. nov., novel eicosapentaenoic-acid-producing marine bacteria isolated from sea-animal intestines. *Int J Syst Evol Microbiol* **53**, 491-499.

Schink, B. (1984). Fermentation of tartate enantiomers by anaerobic bacteria, and description of two new species of strict anaerobes, *Ruminococcus pasteurii* and *Ilyobacter tartaricus*. *Arch Microbiol* **139**, 409-414.

Schink, B. & Pfenning, N. (1982). *Propionigenium modestum* gen. nov. sp. nov. a new strictly anaerobic, nonsporing bacterium growing on succinate. *Arch Microbiol* **133**, 209-216.

Segers, R., Butt, T. M., Keen, J. N., Kerry, B. R. & Peberdy, J. F. (1995). The subtilisins of the invertebrate mycopathogens *Verticillium chlamydosporium* and *Metarhizium anisopliae* are serologically and functionally related. *FEMS Microbiol Lett* **126**, 227-231.

Semple, K. M. & Westlake, D. W. S. (1987). Characterization of iron reducing *Alteromonas putrefaciens* from oil field fluids. *Can J Microbiol* **35**, 925-931.

Shimogaki, H., Takeuchi, K., Nishino, T., Ohdera, M., Kudo, T., Ohba, K., Iwama, M. & Irie, M. (1991). Purification and properties of a novel surface-active agent- and alkaline-resistant protease from *Bacillus* sp. Y. *Agric Biol Chem* **55**, 2251-2258.

Siezen, R. J. & Leunissen, J. A. (1997). Subtilases: the superfamily of subtilisin-like serine proteases. *Protein Sci* **6**, 501-23.

Skerratt, J. H., Bowman, J. P. & Nichols, P. D. (2002). *Shewanella olleyana* sp. nov., a marine species isolated from a temperate estuary which produces high levels of polyunsaturated fatty acids. *Int J Syst Evol Microbiol* **52**, 2101-2106.

Smith, J. E. (1990). Biotechnology. Edited by E. Arnold. Adivision of holdder and stoughton, London

Stackebrandt, E. & Goebel, B. M. (1994). Taxonomic note: a place of DNA-DNA reassociation and 16S rRNA sequence analysis in the present species definition in bacteriology. *Int J Syst Bacteriol* **44**, 846-849.

Stieb, M. & Schink, B. (1984). A new 3-hydroxybutyrate fermemting anaerobe, *Ilyobacter polytropus*, gen. nov. sp. nov., possessing various fermentation pathways. *Arch Microbiol* **140**, 139-146.

Strongin, A. Y., Izotova, L. S., Abramov, Z. T., Gorodetsky, D. I., Ermakova, L. M., Baratova, L. A., Belyanova, L. P. & Stepanov, V. M. (1978). Intracellular serine protease of *Bacillus subtilis*: sequence homology with extracellular subtilisins. *J Bacteriol* **133**, 1401-1411.

Suzuki, M., Taguchi, S., Yamada, S., Kojima, S., Miura, K. I. & Momose, H. (1997). A novel member of the subtilisin-like protease family from *Streptomyces albogriseolus*. *J Bacteriol* **179**, 430-438.

Suzuki, Y., Hatagaki, K. & Oda, H. (1991). A hyperthermostable pullulanase produced by an extreme thermophile, *Bacillus flavocaldarius* KP 1228, and evidence for the proline theory of increasing protein thermostability. *Appl Microbiol Biotechnol* **34**, 707-714.

Van Den Burg, B. (2003). Extremophiles as a source for novel enzymes. *Curr Opin Microbiol* **6**, 213-8.

Van Der Maarel, M. J., Van Der Veen, B., Uitdehaag, J. C., Leemhuis, H. & Dijkhuizen, L. (2002). Properties and applications of starch-converting enzymes of the alpha-amylase family. *J Biotechnol* **94**, 137-55.

Venkateswaran, K., Dollhopf, M. E., Aller, R., Stackebrandt, E. & Nealson, K. H. (1998). *Shewanella amazonensis* sp. nov., a novel metal-reducing facultative anaerobe from Amazonian shelf muds. *Int J Syst Bacteriol* **48**, 965-972.

Venkateswaran, K., Moser, D. P., Dollhopf, M. E., Lies, D. P., Saffarini, D. A., Macgregor, B. J., Ringelberg, D. B., White, D. C., Nishijima, M., Sano, H., Burghardt, J., Stackebrandt, E. & Nealson, K. H. (1999). Polyphasic taxonomy of the genus *Shewanella* and description of *Shewanella oneidensis* sp. nov. *Int J Syst Bacteriol* **49**, 705-724.

Vieille, C. & Zeikus, G. J. (2001). Hyperthermophilic enzymes: sources, uses, and molecular mechanisms for thermostability. *Microbiol Mol Biol Rev* **65**, 1-43.

Watson, J., Matsui, G. Y., Leaphart, A., Wiegel, J., Rainey, F. A. & Lovell, C. R. (2000). Reductively debrominating strains of *Propionigenium maris* from burrows of bromophenol-producing marine infauna. *Int J Syst Evol Microbiol* **50**, 1035-1042.

Watson, R. R. (1976). Substrate specificities of aminopeptidases: a specific method for microbial differentiation. *Methods Microbiol.* **9**, 1-14.

Yanchinski, S. (1985). Setting genes to work: The industrial era of biotechnology. Edited by Viking, London

Yoon, J. H., Kang, K. H., Oh, T. K. & Park, Y. H. (2004). *Shewanella gaetbuli* sp. nov., a slight halophile isolated from a tidal flat in Korea. *Int J Syst Evol Microbiol* **54**, 487-491.

Ziemke, F., Hofle, M. G., Lalucat, J. & Rossello-Mora, R. (1998). Reclassification of *Shewanella putrefaciens* Owen's genomic group II as *Shewanella baltica* sp. nov. *Int J Syst Bacteriol* **48**, 179-186.

Zimmerman, B. K. (1984a). Biofuture: confronting the genetic era. Edited by Plenum press, London

Zimmerman, B. K. (1984b). Trends in biotechnology. Edited by UNIDO, Vienna

Curriculum Vitae

Personal Data	*Nationality* Jordanian *Date of Birth* May/5/1970 *Place of Birth* Amman - Jordan

Experiences

2001 - 2005   <u>Ph.D. student and researcher</u> in the technical microbiology laboratory by Prof. Dr. Dr. h. c. G. Antranikian, Hamburg University of Technology (TUHH), Institute of Technical Microbiology, Hamburg-Germany.

1998 - 2000   <u>Researcher</u> in the environmental engineering laboratory by Prof. Dr. –Ing. N. Räbiger, at Bremen University, Institute of Environmental Process Engineering Technology, Bremen-Germany.

1993 - 1997   <u>Research assistant</u> in biotechnology laboratory by Prof. Dr. I. Mahasneh, at Mu'tah university, Department of Biology,Faculty of Science, Karak-Jordan and at Al al-Bayt university, Department of Biology, Faculty of Art and Science, Mafraq-Jordan.

Education and Academic studies

1994 - 1997   <u>Master of Science (M.Sc.)</u> in Biological sciences at Al al-Bayt university, Department of Biology-Faculty of Art and Science, Mafraq-Jordan.

1990 - 1994   <u>Bachelor of Science (B.Sc.)</u> in Biological Sciences at Mu'tah university, Department of Biology, Faculty of Science, Karak-Jordan.

Patents   Pullulanase (type I) from the psychrophilc bacterial strain *Shewanella arctica* sp. nov.. Registration number 102004046116.3 (Submitted 23.09.2004).

Subtilisin-like serine protease from the psychrophilc bacterial strain *Shewanella arctica* sp. nov.. Registration number 102005028295.4 (Submitted 18.6.2005).

Hamburg, 2006

Acknowledgement

This work is a result of my post-graduate studies at the Institute of technical microbiology, TUHH. After many years, it is impossible to adequately thank all people whose help has made this work possible. I apologize in advance to anyone I forgot to mention in this brief acknowledgement.

First of all, I would like to express my deepest gratitude to Prof. G. Antranikian for his patient guidance and wholehearted support during this work. My gratitude is also extended to Prof. R. Müller for reviewing the manuscript and Prof. F. Keil for taking the chair of doctoral examination committee.

I would like to thank Dr. H. Landsberg and Dr. C. Walbiner from the KAAD for the financial support.

I would like to thank all my colleagues for the friendly working environment and for the help and support. Especially Dr. S. Ressen, Dr. C. Bertoldo, Dr. V. Thiemann, M. Katter, R. Al Khudary, Dr. M. Royter, M. Hess, Dr. R. Grote, D. Biemann, I. Haller, P. Esselun, K. Benz, A. Peters and B. Paella.

I would like to thank my parents, Brothers (Ramzi, Rami and Issa), sisters (Ghada and Ramia) and friends (M. Fliedner) for the support and help in every aspect of my life. Finally I would like to thank the Lord Jesus Christ for being on my side in every step in my life.